Befriending Bumble Bees

A practical guide to raising local bumble bees

Elaine Evans · Ian Burns
Patrick Lhomme · Marla Spivak

Befriending Bumble Bees

A practical guide to raising local bumble bees

by

Elaine Evans, Ian Burns, Patrick Lhomme, and Marla Spivak

Published by:
Wicwas Press LLC
1620 Miller Road
Kalamazoo, MI 49001 USA

www.wicwas.com
Printed in U.S.A.
ISBN:978-1-878075-63-5

CONTENTS

ACKNOWLEDGEMENTS

We would like to thank all the people who have shared their bumble bee rearing experiences with us. We have picked the brains of many fine individuals in the process of developing our methods. We are eternally grateful to Gary Reuter, who provided essential and masterful guidance on our search for a better bumble bee box. A deep thanks to Kirsten and Michael Traynor, who contributed their varied and formidable skills to help this project reach its final stage. We thank Annika Schlesinger, Jeff Hahn, Doug Golick, Tony Ernst, and Michael Traynor for their photography that helps bring the story to life. We thank Michael Otterstatter, Robbin Thorpe, Kristen Murray, Eric Mader, Amy Krueger, and Erik Wivinus for helpful comments and keen editing skills. We would also like to thank the Minnesota Hobby Beekeeping Association for their generous support of our bumble bee research. A special thanks to Paul Metzger for his help, advice, and support.

BUMBLE BEES

The scent of spring flowers wafts through the air to the music of bees buzzing between flowers. A *Bombus bimaculatus* queen forages on rhododendron blossoms. This queen has survived hibernation through winter and now faces the challenge of creating a bumble bee colony. We hope this guide book will give you all the information you need to help her with her challenge.

FOREWORD

One of my favorite books from childhood is *Fortunately, Unfortunately* by Remy Charlip. Pages alternates between fortune and misfortune for the tale's adventurer, showing that if you persevere through challenges, you may come out on the other side with unforeseen benefits.

Since *Befriending Bumble Bees* was first published sixteen years ago in 2007, many fortunate and unfortunate events have impacted bumble bees. Fortunately, people are talking about how important pollinators are. Fortunately, pollinator habitat is being added to gardens, agricultural margins, roadsides, and parks. Every bit of pollinator habitat can make a big difference for the pollinators that depend on the food and shelter there, as well as the many other creatures that depend on those pollinators.

Unfortunately, over the last several decades, pollinators have faced sustained and increased pressures that drive decreases in their populations and prevent recovery of species that are threatened with extinction. Climate impacts have become more obvious with increasing frequency of droughts, floods, fires, and other extreme weather events. Pesticide pressure in both agricultural and urban settings continues to kill pollinators through lethal and sub-lethal actions. Diseases, parasites, and pests can become bigger problems due to decreased ability of pollinators to defend themselves due to poor nutrition and physiological and metabolic pressures from detoxifying pesticides and coping with temperatures outside their tolerances.

Fortunately, increased interest in pollinators has led to more bumble bee research. We've learned a great deal about the impacts of pesticides on bumble bees, the movement of diseases among bees at shared floral resources, and the impacts of climate change on bumble bee ranges and health. Data has been gathered on observed patterns of declines, and on habitat that best supports bees. Researchers have sequenced the genomes of several bumble bees, providing key tools for conservation. We have learned about the minds of the bumble bees with experiments demonstrating their ability to count and play. With improved technology, public participation in bumble bee monitoring has soared. Through public biodiversity portals like bumblebeewatch.org and iNaturalist. org, individuals across the globe have submitted over half a million observations of bumble bees!

Photo: Elaine Evans

INSIDE THE NEST

A peek inside a growing bumble bee nest. The bees produce wax that is used for constructing pots to hold nectar and pollen and also to cover and protect developing young.

Unfortunately, despite these strides forward, we are still missing vital information needed to conserve bumble bees. We still know very little about bumble bee diseases. There are vast gaps in our knowledge of their basic natural history, like their nesting and overwinter habitat preferences, or dispersal distances. To enact relevant recovery plans for endangered species, we need to know which impediments to bumble bee recovery are most important to address.

Fortunately, bumble bee rearing can help us to address some of these key research questions. Working with reared colonies, we can learn about the impacts of diseases, future climate scenarios, pest pressures, and refine our understanding of nutritional needs. These data will help address and mitigate the pressures in certain regions that led to declines of species in the first place. Rearing can be a conservation tool, enabling us to reintroduce species that have been extirpated from particular regions, if population pressures that caused the initial declines in the region have been addressed. Unfortunately, establishing a bumble bee rearing program can be difficult and time consuming. Fortunately, commercially produced colonies can be used for bumble bee research for those without the time or resources to develop their own rearing program. Unfortunately, only a narrow range of species are reared commercially , selection for colonies ready for use in pollination events by rearing facilities may lead to possible changes to bumble bee colony cycles that can affect research results, and there is some lack of transparency about colony treatment during rearing.

Fortunately, bumble bee rearing methods are not mysterious. In this book, we aim to provide the knowledge you need to successfully rear bumble bees along with a little patience, good observation skills, and a bit of creative problem solving. This new edition adds a chapter with rearing methods for cuckoo bumble bees. This group of bumble bees is of particular conservation concern. In North America, half of our cuckoo bumble bees are in decline, and several have not been seen anywhere for decades. We hope that information provided here will aid further exploration of their biology and aid in their conservation.

Let's move from awareness to action. From planting pollinator gardens, to sharing photos of bumble bees, to speaking up about climate change, to observation and study of captive colonies, there are many actions we can take to improve the chances for bumble bees to thrive and for currently endangered populations to recover. Observing captive bumble bee colonies can provide you with a window into a mostly cooperative, sometimes chaotic society with creatures who see the world in a vastly different way than we do. In addition to appreciation for their pollination services that tie together food and ecological systems, let's expand our awareness to include appreciation for their intelligence and creativity and open ourselves to the lessons they can teach us.

Dr. Elaine Evans
University of Minnesota
March 2023

"The real voyage of discovery consists not in seeking new landscapes, but in having new eyes." - Marcel Proust

PART ONE

ABOUT BUMBLE BEES

CHAPTER ONE

BENEFITS OF BUMBLE BEES

Bumble bees are among the most charismatic of insects. Their robust frame and fuzziness combined with their charming habit of buzzing dutifully from flower to flower have brought joy to many. But why would you wish to step beyond these casual glances and actually handle these bees?

Some people raise bumble bees out of curiosity, wanting to see firsthand what happens in a bumble bee nest, as these events are normally hidden from our sight. Others rear bumble bees to learn more about the different species for scientific reasons, as there are vast gaps in our knowledge of the biology of most bumble bee species.

Most people raise bumble bees to benefit from their impressive pollination services, a service that is vital to both food production and the health of our ecosystem. With a large variety of problems afflicting the honey bee industry such as pests, pesticides, diseases, and poor financial returns, it is important to broaden the range of bee species available for commercial pollination.

Unlike honey bees, which were imported to the Americas by European colonists, there are many bumble bees that are native to North America. Having evolved along with our native plants, bumble bees are efficient

Photo: Elaine Evans

GREENHOUSE HEROES

Bumble bees are used to pollinate most tomatoes grown in commercial greenhouses. Bumble bees adapt well to use in green-houses. Commercial bumble bee rearing companies produce bumble bee colonies year-round to serve greenhouse tomato production.

BUZZ POLLINATION

Bumble bees ensure the perfection of tomatoes by performing a special feat called "buzz pollination." The bumble bee grabs the anther cone (the flower part containing the pollen) and shakes it, releasing pollen that would otherwise stay trapped in the anther cone. The pollen is then available to fertilize flowers so that fruit can be produced. Good pollination produces large, evenly shaped, attractive fruits.

and important pollinators of many native wildflowers and crops such as cranberries, blueberries, raspberries, squash, and melons.

One of the qualities that sets bumble bees apart from many other pollinators is a special behavior called "buzz pollination" (see above). In addition, bumble bees pollinate well inside of greenhouses, whereas honey bees are notorious for escaping the greenhouse in search of better prospects outside. The combination of these two characteristics has led to bumble bee colonies being used to pollinate most tomatoes grown in commercial greenhouses. While bumble bee colonies are also used for pollination of field crops, such as apples, cranberries, and raspberries, it has been found that habitat creation near these fields can be just as effective, or sometimes more effective, than field-placed captive-reared colonies.

There are also rare native wildflowers that depend on bumble bees for pollination. As natural habitats become more fragmented, some wild bumble bee populations are in decline and some native wildflowers may lose their pollinators. Without pollination, plants cannot produce seeds. Loss of native pollinators can lead to a decline in native plants, which means less forage for remaining native pollinator species in the coming years, creating a vicious cycle. We can help break this cycle by planting natives in our gardens to ensure local bee populations have ample forage.

Although bumble bees are a common sight in most flower gardens, not all bumble bee species have adapted well to changes that have taken place in the landscape due to the spread of agriculture and urbanization. Bumble bee declines have been documented globally with some species on the brink of extinction. In North America, *Bombus franklini*, a bumble bee native to the western U.S., has not been seen since 2006. Several over species have experienced drastic shrinking of their ranges and decreases in

3

their abundance. The rusty-patched bumble bee, *Bombus affinis*, has been declared an endangered species in the both Canada and U.S.

We need to learn from these tragedies and prevent further bumble bee declines from occurring in North America. We can help reverse this downward trend by learning more about the diversity and biology of local bumble bee populations. By providing suitable nesting and foraging sites, we can support local bumble bee populations. Bumble bee rearing is a great tool for learning more about bumble bee biology and promoting bumble bee conservation.

Bumble bees are creatures of undeniable beauty and charm. You are sure to become fascinated while watching their interactions and marveling at the ever changing sculptures of wax inside their nests. Observing the bumble bees you raised foraging in your fields or gardens will have a deeper meaning when you understand more about these avid pollinators.

BUMBLE BEES AND NATIVE FLOWERS

Photo: Tony Ernst

Here a *Bombus terricola* worker forages on golden rod, *Solidago*, a plant native to Eastern North America. Many native plants benefit from bumble bee pollination. Also, many native plants are important food sources for bumble bees.

CHAPTER TWO

HOW BUMBLE BEES LIVE

When you tell people that you are raising bumble bees, you are bound
to receive some quizzical looks. Can you get honey from them? Do they
sting? Why bother raising bumble bees? While many people are familiar
with honey bees and their amazing behaviors and products, most people
only know what bumble bees look like. They are not aware of how these
insects live.

Bumble bees are closely related to honey bees so the two share many
characteristics. Both belong to the Order Hymenoptera, which also
includes ants, wasps, and other bees. There are about 20,000 bee species
worldwide, belonging to seven different families. Honey bees and bumble
bees both belong to the family Apidae.

Worldwide there are approximately 250 species of bumble bees. Over 40
of these species are native to North America. In contrast, there are only
seven honey bee species worldwide, and no honey bees are native to North
or South America.

Both honey bees and bumble bees live in colonies consisting of a queen
and many workers. They collect pollen and nectar from flowers as their
main food source. To build nests, both honey bees and bumble bees
produce wax from special glands on their abdomens.

One prominent difference between honey bees and bumble bees is that
honey bees collect large amounts of nectar from flowers and convert it to
honey inside the nest. These honey stores sustain the honey bee colony
through the winter months. Bumble bees do not actively evaporate water
from nectar to turn it into honey but store nectar directly into wax pots,
collecting only enough to last through a few days of bad weather. As their
colonies typically do not survive the winter, bumble bees have no need to
store large amounts of nectar. This means there is nothing there for us to
take for our own use. The primary benefit bumble bees give people is their
pollination services.

Bumble bees can sting, but their stinger is not barbed like the stinger of
honey bees. This means that unlike honey bees, bumble bees can sting

Photo: Ian Burns

QUEENS *vs.* WORKERS

Here you can see the dramatic size difference between queens and workers as the queen walks up behind a lightly colored newly emerged worker. Bumble bees do not grow larger once they emerge from their pupal cocoons. Bumble bees do all of their growing as larvae. Workers vary greatly in size depending on how much food they consume as larvae and can be much larger than those pictured here.

repeatedly with their straight stinger. Since the stinger remains intact, they do not make a fatal sacrifice with each sting like their cousins the honey bees, whose barbed stinger remains behind. However, most bumble bees are docile, stinging only in defense when their nest is disturbed. A few bumble bee species can be quite aggressive in protecting their nests.

Honey bees form wax into uniform hexagonal cells, which many cultures use as a symbol of stability. In sharp contrast, wax is formed into pots of varying shape and size in a bumble bee colony. Bumble bee nests constantly change as the colony grows, giving a chaotic appearance to the hive.

Not only are their homes more organized, but honey bees also have more complex forms of communication. Many people know about the dance language that honey bees use to communicate with each other. Bumble bees do not have any such form of communication. However, there are other ways that bumble bees exchange information. Like many other animals, they use chemical signals, called pheromones, to communicate with each other. In some bumble bees, returning foragers run excitedly in random patterns, possibly in an attempt to stimulate other bees to forage.

Bumble Bee Development

We are most familiar with the fuzzy adult bumble bees we see visiting flowers in the summer, but each bumble bee spends a substantial portion of its life in other developmental forms. We don't usually see these different forms, as they remain concealed within the shelter of the hidden nests.

Bumble bees go through four developmental stages: egg, larva, pupa, and adult. The queen will start a colony by laying a clutch of approximately eight to fourteen eggs. The eggs are white and sausage shaped, measuring about one-eighth of an inch in length. They will remain in this egg stage for about five days. Upon hatching, the larvae appear similar to the eggs.

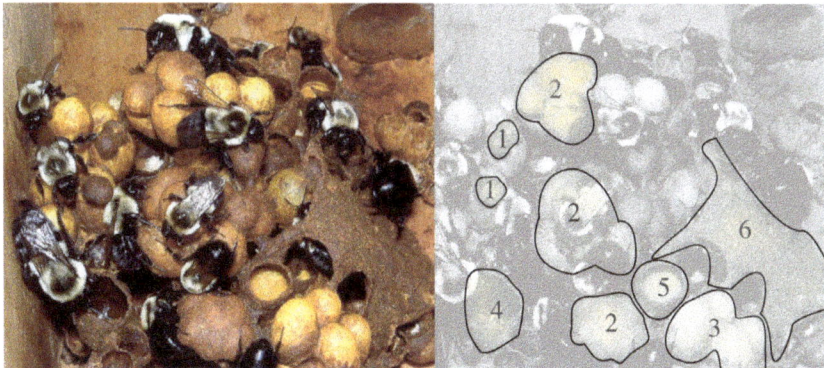

Photo: Elaine Evans

COLONY DEVELOPMENT

The *Bombus impatiens* colony shown here has all stages of bumble bee development present. Bumble bees sometimes build a wax cover over the entire nest, which can help regulate nest temperature. 1. Eggs 2. Larvae 3. Pupae 4. Nectar pots 5. Pollen pots 6. Wax cover

BUMBLE BEE EGGS

Here the wax cover has been removed to reveal the eggs underneath. Bumble bee eggs are about one-eighth of an inch long, clearly visible to most unaided eyes, once the wax cover is removed. When covered, you will only see a small wax lump.

Photo: Annika Schlesinger

Close examination will reveal that each larva has a mouth and can move. The larvae develop rapidly in size. The larvae remain hidden underneath a wax covering at this time, but you can observe these wax covered clumps expanding in size as the larvae grow. Wax is added to the clumps to ensure the larvae remain sealed.

Under normal conditions the larval stage lasts about fourteen days. When temperatures sink or high quality pollen is not available, development can take longer. The larvae go through four stages, known as "instars," where they grow and then molt, changing from pale colored to pinkish or brown. The larvae eat a pollen and nectar mixture, provided by the queen and/or workers. This protein (pollen) and carbohydrate (nectar) rich food enable their rapid growth.

BUMBLE BEE LARVAE

Photo: Doug Golick

The bumble bee larval stage lasts approximately fourteen days. The larvae usually grow under a wax covering. Here the wax has been removed to reveal the larvae.

Bumble bees use two different methods to feed their larvae. In some species, returning foragers bring pollen directly to clumps of larvae and deposit it into pockets at the base of the wax cells. The larvae feed themselves from the pockets, which is why these bees are called pocket-makers.

Other bumble bee species store pollen in separate cells. Bees

8

BUMBLE BEE PUPA

During the pupal stage, an amazing transformation takes place. The cells of the legless, wingless larva are rearranged to form the body of the adult bee. The pupa remains sealed in a silk cocoon and does not eat. Here the cocoon has been partially removed to reveal the pupa.

Photo: Doug Golick

within the nest then transfer the stored pollen and feed it to the larvae by opening their wax covering. These bees are called pollen-storers.

It should be noted, however, that when colonies of pocket-making species become large, they too may begin to store pollen in cells, thus blurring the distinction between pocket-makers and pollen-storers. The distinction between these two methods of brood feeding is important for bumble bee rearing, as bees that begin brood rearing as pollen-storers are easier to rear in captivity.

After several weeks of feeding, larvae spin individual cocoons, much like caterpillars, using special silk producing glands found in their mouths. These cocoons are formed under the wax covering. After spinning their cocoon, the bumble bees are called pupae, an intermediate stage between larva and adult.

Once a cocoon is formed, the wax is often removed and reused by workers, revealing the silk cocoon underneath. Inside the cocoon, an

EMERGING ADULT

This fully formed adult uses its strong mandibles to chew its way out of the pupal cocoon. The bees are pale colored when they first emerge. After several hours, they will be fully pigmented.

Photo: Annika Schlesinger

amazing transformation takes place. Cells within the body of the bee rearrange themselves and the legless larva develops legs and wings to enable the adult bee to walk and fly. Large eyes form on the head that will permit the adult bee to adeptly navigate the world outside the nest. A long tongue forms so that the adult can sip nectar from flowers, collecting this nectar to share with her nest mates. After about fourteen days, the adult bee emerges from the pupal cocoon. The once wingless, legless larval eating machine has transformed into a fuzzy, winged adult with six legs, capable of flying miles from home to collect nectar and pollen.

Typically, it will take five weeks from the time the queen lays an egg until an adult emerges. However, development time is affected by temperature as well as the quality and quantity of food available, and as such, typically ranges from three to seven weeks. Worker bees usually live an additional three weeks as an adult after emergence from the pupal cocoon.

Life Cycle of a Bumble Bee Colony

Bumble bees are social insects, which means that many bees, all biologically related to one another, live together, forming a colony. As in most societies, through cooperation and division of labor, the social unit has greater productivity than the sum of each individual's efforts. The "personality" of the colony grows out of the collective actions of the individual residents. Although at first it appears that chaos reigns in a bumble bee colony, there are divisions of labor and rules of organization. Colony members are divided into three different castes, each with specialized duties: queen, worker, and male.

After the queen has founded a colony and her first clutch of workers emerges to help her in providing for the hive, the queen's primary job is to lay eggs. The workers gather food, care for the young, and clean and defend the nest. The males' purpose is to leave the nest to mate with virgin queens from other nests, ensuring future genetic diversity.

In bumble bees and all other insects in the Order Hymenoptera, unfertilized eggs become males, and fertilized eggs become females. The queen is the only female that will mate; only she can produce female offspring by adding the sperm stored during her mating to fertilize her eggs. All fertilized eggs are females and will either become workers or queens, depending on how they are raised and fed.

The queen produces males when she lays unfertilized eggs. Barring the odd accident, males are produced late in the colony life cycle. Workers, though they never mate, are also capable of producing offspring. This ability of workers to produce their own young often leads to conflicts between workers and the queen over whose eggs will be raised by the colony. However, since the workers have never mated, they are incapable of producing eggs that will become other workers or queens. All of their eggs lack fertilization and so are destined to become males.

Queens mate once they emerge as young adults from the nest where they were born and raised. Mating usually occurs in summer or early fall. This is the only time period during which the queen will mate. After a queen mates successfully, she stores the sperm in a structure in her abdomen called a spermatheca. Some bumble bee queens mate with several males, mixing the sperm in the spermatheca. The sperm stored from this brief mating period is used throughout the queen's life, which usually lasts about one year.

After mating, the queen enters diapause, a form of hibernation. She will find a suitable place and dig herself into the ground, surviving underground through even the coldest months of winter. Bumble bee queens have a chemical in their blood that is equivalent to antifreeze, preventing their cells from freezing in very low temperatures.

Lying dormant under the often frozen ground, these queens with sperm stored in their spermatheca are the only connection to next summer's buzzing bumble bees. Without them, bumble bees would cease to exist, as these mated queens are the only ones who survive the winter.

In the spring, the queen emerges from the ground when soil and ambient temperatures are warm enough for her particular bumble bee species. See the table in Chapter Ten for typical emergence times for North American bumble bee species.

After emerging, the queen searches for a good nesting site. You can see queens darting low to the ground, investigating potential nest sites. Typical nest sites are small holes in the ground or above the ground in piles of grass, with nest preference varying among species. Bumble bees have been discovered nesting in abandoned bird and rodent nests, rolled up carpets, insulation, underground cavities, wood piles, clumps of grass, and underneath sidewalks. Queens will spend many hours looking for the perfect spot.

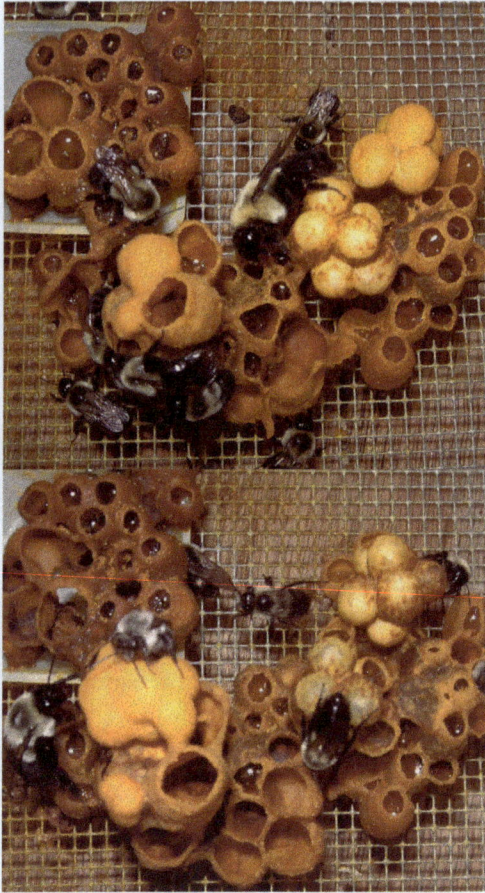

Photo: Ian Burns

COLONY PROGRESS

Bumble bee colonies are in a constant state of flux. Tiny changes to the shape and size of structures in the colony are being made throughout the day. The pictures here show the changes in one colony over just a few days. The larvae in the top right corner of the top picture become pupae in the bottom picture. The small clump of larvae midway along the left side of the colony becomes a large clump of larvae in the bottom picture as the larvae rapidly grow. Nectar pots are expanded or demolished or simply altered to compensate for the constantly shifting architecture of the nest.

After choosing a suitable location, the queen's work has just begun. Now the queen must find food to support the development of her ovaries before she can begin to lay eggs. She also needs to produce wax, so she can build the pot to hold the nectar she will collect. The independent queen forms the foundation of the colony and provides for its success.

Once she has decided on a nesting spot, the queen will forage for nectar and pollen. Inside the nest, she will build a wax pot in which to store some gathered nectar. She will form the pollen she collected into a ball, then lay a clutch of eggs (usually eight to fourteen eggs per clutch) on this pollen ball, and cover the eggs with wax.

The larvae will develop in a clump under the wax, eventually separating into their own wax cells. Finally the larvae will form their own cocoons, pupate, and emerge as fully formed adults. The queen lays a second batch of eggs after the first batch has pupated.

When the queen lays eggs, they may or may not be fertilized by stored sperm. As mentioned previously, all fertilized eggs laid by the queen develop into female bees. Unfertilized eggs always develop into males. Unlike the females, who can become either workers or queens, the males only have one caste.

The larvae are in effect just one big stomach that is in constant need of food to grow and develop. After the first clutch of brood hatch, the queen must continue to forage to supply her young with pollen and nectar. When she is not searching for food, the queen spends time incubating eggs and larvae to keep them warm.

When the second clutch pupates, the queen switches to continuous egg laying. With her young worker force to aid in the care of her offspring, the queen is now freed from the task of feeding the larvae. She usually stops foraging at this time, allowing her daughters to carry out the task. The queen concentrates her time and energies on laying eggs into wax, cup-shaped egg cells constructed by herself or the workers.

The colony continues to grow during the summer. At the peak of the colony's life cycle, populations range from a few dozen to a few thousand bumble bees. *Bombus impatiens*, a commonly reared species in the U.S., typically has 200 to 500 workers residing in the colony at its peak.

Successful colonies will produce a new generation of queens. Queens and workers are both female and develop from fertilized eggs, but queen

EGG CELL

A queen sits above an egg cell, constructed of wax, in which she will lay eggs. The eggs are then covered with wax. This egg cell has been made on top of some pupae.

Photo: Annika Schlesinger

production requires a greater quantity of food than worker production. Queens are normally produced later in the season after the colony has a sufficient number of workers to provide the copious amounts of food the hungry queen-destined larvae require.

Seeing queen production in a colony is an indication that the colony is successful. However, it also indicates that the colony will soon disintegrate. Typically, once queen production begins, subsequent females produced by the colony become queens. Although queens may do some foraging, they do not fulfill the work duties of the workers. In addition, the queen and sometimes the workers, will begin to lay unfertilized eggs, which become males.

For most colonies, the production of males and new queens signals the end of the colony life cycle. Bumble bee colonies in temperate regions do not usually survive more than one summer. There are reports of some tropical bumble bee colonies continuing for over a year.

These newly produced males and queens will leave their home nest to mate, helping to ensure greater genetic diversity. After successfully mating, the newly mated queens search for a suitable location in which to spend the winter. The mated queens are the only colony members that survive the harshness of winter. The remaining males and workers die off in late summer or fall, and the nest is abandoned.

The entire cycle starts again the following spring when the mated queens from the previous fall emerge from hibernation and begin to form their own colonies. As a bumble bee rearer, the first sign of spring will no longer be the return of migrating birds or the flush of green returning to the landscape. Spring will be announced by the loud hum of a queen zooming past on her mission to found a new colony.

Photo: Elaine Evans

GROWTH OF A BUMBLE BEE COLONY

Stage 1: Colony Initiation. Eggs and young larvae (first brood).
Stage 2: Multiple Generations. Adults (first brood), pupae (second brood), larvae (third brood), and eggs (fourth brood).
Stage 3: Continual worker production.
Stage 4: Reproductive Phase. New queens and males.

PART TWO

REARING BUMBLE BEES

RAISING BUMBLE BEES

Step-by-step Summary of How to Rear Bumble Bees in Captivity

1.   Catch a local, mated queen.

2.   Place the queen in a small box with a pollen ball and nectar. Replace the pollen ball every few days if there is no sign of egg laying on the pollen ball.

3.   Once there are eggs, continue to provide fresh pollen every few days so the queen can feed the larvae.

4.   When workers emerge, transfer (or transform) the colony to a larger box. Continue to supply pollen and nectar.

5.   Once colony population exceeds twenty, the colony can be opened to the outside.

CHAPTER THREE

WHICH BEES ARE BEST FOR DOMESTICATION?

With over 250 species of bumble bees worldwide and over 40 in the United States, there are many bumble bees from which to choose. However, some bumble bees are more difficult to rear. All of our rearing experiences are with species local to the north central United States. We have had the best success rearing *Bombus impatiens*. We also have had success rearing *Bombus bimaculatus* and *Bombus griseocollis*. We also raised *Bombus affinis* when it was a commonly found bee. We have not had success rearing *Bombus auricomus* or *Bombus fervidus*, but others have successfully reared these species. There is still much room for experimentation with new techniques in bumble bee rearing. Perhaps you will meet with success where others have failed.

The bumble bee species you choose to rear depends on your purpose. If you are interested in learning more about their biology, or simply curious to see first hand what happens within the nest, then nearly any species can be used. If pollination is your goal, you will want to make sure you choose a species that has large colonies and is easy to rear. See table in Chapter Eleven for a list of North American species and their suitability for rearing.

POCKET-MAKERS *vs.* POLLEN-STORERS

Pocket-makers feed their larvae by depositing pollen in an opening at the base of the clump of wax covered larvae. All the larvae in the clump feed on this pollen.

For pollen-storers, the wax covering the larvae is opened, and each larva is fed individually, then the wax is resealed. The pollen-storers seem more adaptable to artificial rearing than their counterparts, the pollen pocket-makers. One reason may be that pollen-storers adjust to you providing them with pollen as they are accustomed to storing it, and then transferring it from the storage cells to the larvae.

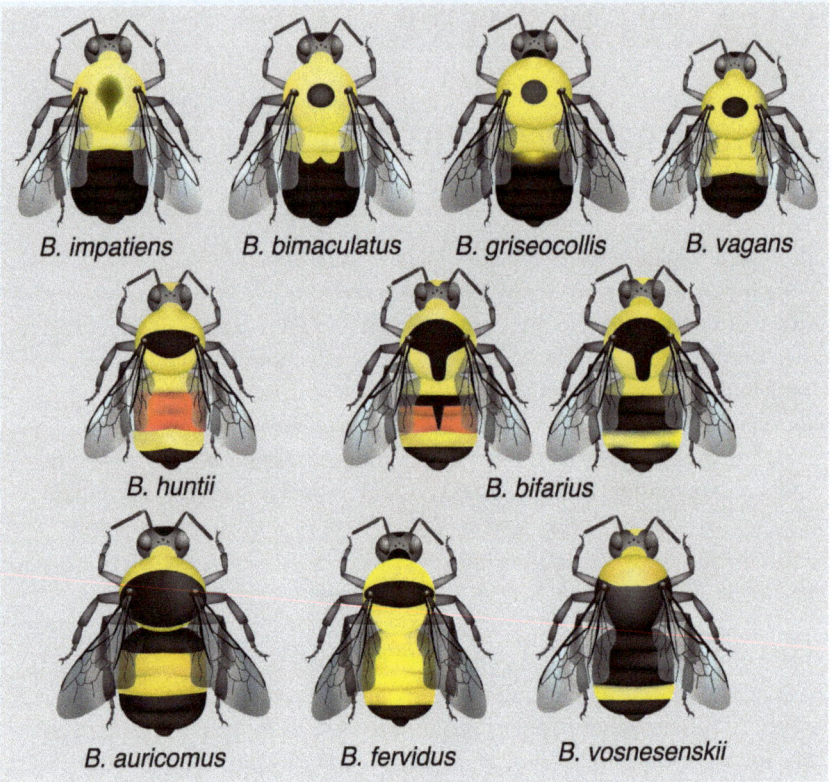

B. impatiens B. bimaculatus B. griseocollis B. vagans

B. huntii B. bifarius

B. auricomus B. fervidus B. vosnesenskii

Illustration: Elaine Evans

COMMON BUMBLE BEES OF NORTH AMERICA

Some common North American bumble bees queens: Queen bumble bees look much the same as workers. The main difference is that queens are larger. *Bombus impatiens* is the most commonly reared bumble bee in eastern North America. *Bombus vosnenskii* and *Bombus bifarius* are good candidates for rearing among western bumble bees.

It is extremely important that you take the time and effort to learn how to identify your local bumble bee species. Some bumble bees are rare and should not be collected. *Bombus affinis*, *Bombus occidentalis*, *Bombus terricola*, *Bombus franklini*, and *Bombus pensylvanicus* populations have dropped in recent years. We need to learn better ways to protect our bumble bees so that rare species are not driven to extinction. If you are lucky enough to find a rare bee, please leave it in the wild.

Although the task of trying to identify bumble bee species as you catch a quick glimpse of the insect darting between flowers may seem daunting at first, many bumble bees species can be distinguished based on their color patterns without the aid of a microscope. With a little practice, you will be able to identify many of them as they take a brief stop on a flower. Some species will require closer examination.

You can capture bees directly off flowers into a clear plastic jar that will allow you to examine the bee. To slow bees down so you can get a good look at them, cool the captured insect in a refrigerator or portable ice chest. Since bees are cold blooded, cooling immobilizes them, allowing you to examine them closely. The bees will warm up relatively quickly and fly off, unharmed. You can take photographs or make sketches to serve as reminders.

After you spend some time looking closely at the bees you too will learn to distinguish many of the bumble bees on blossoms or captured in your glass jar. See the Bumble Bee Identification for North America section in References for a list of excellent guides to identifying bumble bees.

IMPORTANT

It is important that you capture your queens within the same region you plan to raise them. When you transport bees into different areas, you run the risk of also transporting diseases or parasites for which local bees may not have any resistance. This puts the entire local bee population at risk. By keeping things local you are also ensuring that the bees are properly adapted to your climate and vegetation.

We recommend that bumble bees being mass-reared and transported are carefully monitored for signs of disease and records are made available to ensure the ability to trace disease should there be any outbreaks of disease within rearing facilities.

CHAPTER FOUR

CATCHING QUEENS

To begin rearing bumble bees, the most important things you will need are mated queen bees that are capable of producing an entire colony of bees. You cannot initiate a colony with a worker bee, as she is only capable of producing male offspring. Some people set out empty boxes with the hope that queens will decide to enter on their own and nest in them. Even with hundreds of boxes, the success rates reported are typically very small and generally not worth the effort. Although it may take a while before you locate good spots for capturing spring queens, finding them on flowers is the most reliable method for locating them.

Distinguishing Bumble Bees from Other Insects

Of course, the first step in collecting bumble bee queens is making certain the insect you are pursuing is indeed a bumble bee. For most people the name bumble bee conjures up an image of a big, fuzzy, yellow and black insect. However, not all big, fuzzy, yellow and black insects are bumble bees.

Because bumble bees have a sting to defend themselves, many animals have learned to avoid them. Other insects mimic the shape and coloring of bumble bees to gain protection from potential predators. Even though these other insects don't have a sting, many predators leave them alone, simply because they look like something that could sting. Among these impersonators are flies, moth. Other bees and wasps can also be confused with bumble bees.

To avoid collecting mimics, closely examine the insects you catch. Make sure that they have two pairs of wings, front wings, and hind wings on each side. This inspection will eliminate the possibility of collecting a fly, as flies only have one pair of wings. In addition, flies often have very large eyes and oddly shaped mouth parts. Many do not have the constricted waist characteristics of the bees. To avoid collecting a moth, look closely at the tongue of the insect when it is not feeding. Moths coil their long tongue into a roll, whereas bees fold their tongue and tuck it under the mouth. Some solitary bees appear similar to bumble bees. Most

Photo: Jeff Hahn

BUMBLE BEE MIMICS

Here are two insects that resemble bumble bees. Both these insects are flies. The flower fly (Syrphidae) on the right can be seen acting much like a bumble bee, visiting flowers. The one on the left (*Laphria* spp.) actually captures bumble bees in flight and eats them.

solitary bees will be much smaller than bumble bee queens. Also, the hind legs of solitary bees will be different. Bumble bees have a clearly visible concave area on the third pair of legs (hind legs), called a "pollen basket," for collecting large balls of pollen. If the pollen basket is full of pollen, you will see the brightly colored pollen attached to their hind legs. Some solitary bees collect pollen on their hind legs but if you look closely you will see that this pollen is sticking to brushes of hair on the legs, rather than the pollen being packed into a pollen basket.

Once you become familiar with the color patterns of your local bumble bees, you will recognize that another insect you capture does not match any of these patterns. When collecting bumble bees for rearing purposes, keep in mind that only the queens should be collected; they are the only bumble bees capable of founding a colony. Most other bees and flies that might end up in your insect net will not be as large as the queen bumble bees.

Is the Bumble Bee You Caught a Queen?

Typically, queens are noticeably larger than worker and male bees. However, there are many bumble bee species. Some are larger in size than others, and there is size overlap between a queen bee of a smaller species and a worker of a larger species. In addition, bumble bee size varies with the amount and quality of food consumed by the developing bee. Due to this size range, it is important to become familiar with the bumble bee species local to your area.

The timing of your search is key. Queens are the only bumble bees that hibernate through the winter. You need to search for them after they have emerged from hibernation and before they start a nest on their own. Different species have different emergence times. If you search in early spring, most of the bumble bees you see will be queens. Make

Photo: Michael J. Traynor

SPRING QUEENS

A spring queen sips *Rhododendron* nectar in the warm spring sun on a perfect bumble bee queen hunting day. Keep a notebook with your collecting supplies. Note dates and locations of flower blooms used by spring queens and the bee species found there.

observations and take notes on dates of the emergence of bumble bee queens in your area. Although the times vary greatly depending on the weather, after a few years you will have a general time frame in which you can expect to see queens.

Success is more likely with early queen capture. Workers of the early emerging species will be out foraging at the same time as queens of late emerging species. However, these workers will all be noticeably smaller than the queens. Queens with established nests will have pollen loads; thus it is best to capture bees with no pollen loads. Otherwise you may capture a queen who has already started a colony, leaving her offspring abandoned.

Capturing males by mistake should not prove a problem. Males are not usually present until later in the season and they are roughly the same size as workers. Their abdomen is slightly longer than that of workers, as are their antennae. They will often have a patch of yellow hair on the front of their head.

Where to Catch Queens

The best place to catch queens is at their favorite flower, while they are preoccupied with gathering nectar. They will stop and feed at the flower for a moment, giving you a chance to capture them. Queens may also be found flying low to the ground darting back and forth as they search for a nesting place. However, it is more difficult to catch them when they are searching for nests, because they are more likely to dart off and quickly disappear.

See the table in Chapter Eleven for a list of early spring blooming flowers preferred by bumble bees. Queens can pop up anywhere: in your backyard, along the boulevard as you walk to work or the store, on the dandelions in the park. You know you are a committed bumble bee rearer when you find yourself bringing an insect net and some vials with you everywhere you go between the months of March and June.

How to Catch Queens

There are two basic approaches for catching foraging queens. One technique is to sneak up with a small jar with ventilation holes and quickly place it over the queen while she is distracted on a flower. We use jars that have a diameter of about one inch and are two inches tall. Jars work best if

JAR METHOD

One of the easiest ways to capture a queen is while she is preoccupied. As she forages on a flower, simply scoop her into a jar and quickly fasten the lid in place. Be careful not to catch her legs in the lid.

Photo: Michael J. Traynor

there are many bees foraging at one patch, as you can quickly slip a lid on the jar and move on to the next bee, without spending time transferring the queen from net to jar.

The other approach is to simply make a quick swoop with your insect net and either flip the end of the net around so she is sealed in the closed end of the net, or gently slap the net onto the ground. A net works better than a jar if the bees are flighty and difficult to approach or visiting taller plants.

A good insect net will be light and easy to handle. Make sure that the net is doubled over after capturing the queen, and that there are no holes in the netting. It is very frustrating to catch a queen in your net and then watch her escape. Transfer the queen quickly from the net to a jar. Queen bumble bees have powerful mandibles and can chew through netting if confined for a long time.

Photo: Michael J. Traynor

A QUICK SWING

When you see a queen, swoop her up in your net and with a quick turn of the wrist, flip the net over on itself, so the opening is sealed and the queen cannot escape. Another option is to gently slap the net opening onto the ground. Either way the queen will be captured in the net and you can easily move her into a small jar for transport home.

Queens can be kept in vials or jars for transport

home. To transfer the queen from the net into the jar, isolate the queen in the tip of the net holding the tip up in the air. Bumble bees will usually move upward when given the chance. Then bring the jar into the net from underneath and gently maneuver the queen into the jar. Although this sounds complex, it is very simple in practice.

Don't be afraid of being stung by the bee while trying to maneuver her into the jar. When the queens are stuck in the net, their main priority seems to be escaping out of the net. The best mode of action is to try and transfer her into the jar as quickly as possible, or she will find a way out and zip away.

Carefully hold the queen in one end of the net by folding it over on itself. Place the jar inside the net and work it up toward the queen. Once the queen is in the jar, bring the lid inside the net with your free hand and slip it over the jar. Be sure her legs are not at the rim as you slip the top into place. Complete the transfer in the net, so that she will still be in the net should she manage to escape out of the jar. Make sure the jar contains

TRANSFERRING FROM NET TO JAR

First isolate the queen in the net. Simply hold the tip up, and she will usually move up. Fold the tip over so she cannot escape back down.

Slip a small jar into the net. Gently place the jar over the queen. Hold the jar so its opening is against the netting, preventing the queen from leaving the jar.

Once the queen is in the jar, slip a lid into the net and over the opening of the jar. Make sure you do not catch the queen's legs on the rim of the jar. You have now successfully captured a queen!

Photo: Michael J. Traynor

27

ventilation holes large enough to allow air in, but small enough to keep the bees from escaping.

Put only one queen in each jar. Keep the jars out of the sun to avoid over-heating the bees. Remember that in the sun your car is like an oven. You can put them in a cooler with ice if you are in warmer weather, but be sure that the queens in the vials remain dry and are not in the same area as melted ice water. Try to get the bees to their housing in a controlled environment within a few hours of collecting them.

Searching for spring queens is one of the most enjoyable aspects of bumble bee rearing. Wait for a warm, sunny spring day, grab your net and your jars and head out in search of blossoms and the queens you will find there, attracted by the sights and scents.

CHAPTER FIVE

YOUR QUEEN'S NEW HOME

Now that you have your queen, you will need to place her in an environment that will encourage her to begin laying eggs. There are many different housing designs that are suitable for raising bumble bees. When initiating nests, queens seem to prefer a relatively small space. This is why we like to house the queens in a "starter box." As the colony grows, it will require more space. Starter boxes can have a removable wall to expand to a larger size when necessary. If you want less than a dozen colonies, boxes with expandability may be the best choice. However, if you want to produce many colonies, smaller starter boxes will allow you to fit many more queens onto your shelves.

Success rates in starting a colony from a wild captured queen may vary from 20% to 50%. You should always plan on housing at least twice as many queens as your desired final number of colonies. Greater success rates can be obtained using queens that were themselves raised in captive colonies. If your queen does not start laying eggs within two weeks of capture, she is unlikely to do so.

Boxes can be made of any material that is hard enough to resist the strong mandibles of the bumble bees. Plastic, wood, Plexiglas®, heavy cardboard and poured plaster are also possible construction materials. If reusing nesting boxes, sterilizing is important.

STARTER BOX

This is our preferred starter box because of the ease of accessing pollen through the sliding door, and monitoring and replenishing the nectar in the feeder, which is accessed by the bees through the floor. The design is available at befriendingbumblebees.com

Photo: Michael J. Traynor

Several different nest box designs have been described by bumble bee researchers and enthusiasts who have successfully raised queens (see Nest Box Designs in References). Below follows a nest box design that has worked well for us. Our system may not be the most efficient for you. Feel free to experiment to find what works best for you.

Starter Box

After catching a suitable queen bumble bee, we place her in a small starter box to encourage her to lay her first clutch of eggs. This is a separate, non-expandable starter box, whose dimensions are 2.5 x 5 x 3.5 inches.

The top and bottom of the box are hardware cloth, a stiff metal mesh screen available from hardware stores. Be sure to ask for one-eighth inch hardware cloth, which has eight squares per inch. The hardware cloth floor allows debris to fall through, but keeps the bees in and provides ventilation.

The sides are wood, except for one side which has a sliding metal door that can be raised to provide access to the inside of the box. There is an inner clear plastic door or window that allows you to look inside the nesting chamber to see if the pollen ball needs to be changed. You have easy access to change the pollen ball, without letting the queen escape. The interior space is divided into two sections; a nesting area and a food

TRANSFER THE QUEEN TO STARTER BOX

To release the queen into the starter box, simply raise the door, remove the lid from your jar and place the open end of the jar into the starter box.

Photo: Michael J. Traynor

If the queen refuses to leave the jar on her own you can persuade her with some light jiggling. Once she is in the box, remove the jar and slide the door down. You need only open the door a crack to slide the pollen ball on its plastic dish into the box.

area, separated by a piece of wood with a hole large enough for the queen to move between sections. The internal dimensions of the nesting chamber used for the queen are 1.75 x 1.75 x 2 inches.

Nectar substitute is accessible through a plastic container set below the hardware cloth floor. Cotton dental wicks, available through dental supply catalogues, are used to wick the nectar substitute up to the area where the queen lives. You can also place a small jar of nectar with small holes in the lid directly onto the hardware cloth roof.

We supply the queen with a pea sized pollen ball coated with wax on a small plastic dish in the nesting side of the box. Methods for feeding your bees are described in detail in the Chapter Six.

Finishing Box

Once the first clutch of workers has emerged as adults, we move the bees and the brood into a larger box. The larger box we use is 9 x 12 x 7 inches. The transfer is easiest if the queen has laid her eggs on the pollen ball or at least somewhere in the pollen dish. Sometimes they will lay eggs elsewhere making it difficult to move the brood without causing damage.

Bees can be picked up by their back legs using blunt ended tweezers and dropped into the larger box. The most important bee is the queen, so be sure to treat her gently. Without the queen, no colony can continue to grow.

Photo: Michael J. Traynor

FINISHING BOX

A larger box is used to house the colony after they outgrow the starter box. Give the colony lots of room to grow as the season progresses. This box has a tube attached to the entrance that can be used if the colony is housed indoors and needs to be connected to the outside.

WORKING WITH MATURE COLONIES

A piece of clear plastic with a hole in the center that fits over your finishing box will make it easier for you to feed and observe your mature colonies.

Photo: Elaine Evans

To help keep things clean, you can add a false bottom of hardware cloth one inch from the bottom of the larger box to let debris fall through, or you can add a smaller second room which the bees can use for relieving themselves. Neither of these measures are needed if you intend to let the bumble bees forage freely as they will defecate outside the nest.

Your colonies will need a constant supply of food. Provide pollen (an amount equal to about one-third of the size of the brood) and a nectar substitute daily. For nectar, a feeder can be placed directly in the nest box but must be monitored and refilled when necessary. See Chapter Six for more information on feeding methods.

Some of our boxes have clear plastic sides or tops to facilitate observation of bumble bee behavior. It is useful to have at least one extra cover that is made of clear plastic with a hole in the center. The hole should be large enough for your hand to fit through. By using a solid sheet of plastic in combination with the cut sheet, you can gain access to the colony without letting bees escape. Place the solid piece of plastic on top of the sheet with the hole. When you see that no bees are flying or moving near the top, slide the solid piece off, giving you access to the hive. Once you are done examining the nest, you can use the plastic sheets again for the next colony.

If your colony is to be placed outside to forage freely, all sides should be wood, or some other opaque material, so that the nest stays dark. We recommend making one or two small (1" diameter) screen-covered ventilation holes anywhere on the box. If it becomes necessary to provide nectar to your bees, you can place a feeder within the box. Nest boxes can also be buried underground or in sand with an entrance tube connecting the colony to the outside. This can help the bees maintain favorable temperatures and humidity in their nests. It also will help conceal the nest from predators.

Environment

Bumble bee colonies are usually raised in a dark environment with 50% humidity, at temperatures above 70° F. Our experiences support the recommendations, although they might not all be necessary for all bumble bee species. There are logical reasons why these conditions prove helpful for bumble bee colonies. The humidity keeps the pollen and nectar from drying out. The darkness mimics lighting conditions in underground nests and helps the bees remain calm.

A red light is useful to observe the bees, as bees do not see in the red color range, and so the bees fly much less under these lighting conditions. However, if you have white lights on for a brief period each day, it will not disturb the bees. Some bumble bees may have greater success with regular periods of light and dark.

The higher temperatures ensure that the brood is kept warm and may reduce the time and energy the queen spends incubating brood. Alternatively, though more elaborate, you can put heating units on each box. This way you will not have to spend your time in a hot bee room. If you cannot keep the room warm, be sure that you provide the queen with insulating material. Upholsterer's cotton is an ideal material. Do not substitute cotton balls as the bees will become tangled in the fibers. Any colony placed outdoors should have insulation. However, if your goal is to observe interactions within the nest, this insulating layer will make it difficult to see what the bees are doing.

CHAPTER SIX

FEEDING YOUR BEES

Bumble bees are use pollen, and the microbes that live in the pollen, as their source of protein and nectar as their source of carbohydrates. Queens need pollen and nectar for their own sustenance, to support egg production, and to feed their developing brood. Once the queen is confined within her new home, you must provide her with a constant source of pollen and nectar.

Be sure to keep things fresh and clean. Wash your hands before handling the pollen or wear disposable gloves. Make sure any utensils or containers you use are clean.

Your methods for feeding will depend on how many queens you are raising. If you have ten queens, it is no problem to go into each box every other day to make sure they have access to nectar substitute and to give them fresh pollen. If you have one hundred queens, this becomes a time consuming task and so you will want to develop a more efficient method.

Pollen

You can use pollen that has been collected by honey bees. This pollen should be as fresh as possible; ideally it should have been frozen as soon as possible after collection from the honey bees and used within a month or two. Contact a local honey bee beekeeper to see if you can buy fresh pollen from them. There is ongoing research into possible substitutes for honey bee pollen, but currently, fresh honey bee pollen is the best available source of pollen for bumble bee colonies.

One-half pound of pollen per colony should be more than enough to raise them to the point where they can forage independently, if you do not intend to keep them captive. If the pollen is fresh, you will be able to take a pollen load and crush it easily between your fingers. Dried pollen is not a good substitute for fresh pollen as many important nutrients are lost. Sort through the pollen by hand or with a sieve and pick out any debris.

FINDING POLLEN

To find fresh pollen, contact your local beekeeping association. They can help you find someone in your region who sells fresh pollen. It is best if they freeze the pollen as soon as possible after collection. If you need to have it sent through the mail, have it packed with cold packs and sent as quickly as possible.

Photo: Michael J. Traynor

Mix fresh honey bee pollen with enough sugar solution (see the nectar substitute recipe on page 38) to form a stiff dough. Roll this dough into a cylinder shape with a diameter of approximately one-half of an inch. To make the pollen balls for the queen, cut the cylinder into one-quarter of an inch slices and form into balls. One-quarter of a cup of pollen will make about twenty-five pollen balls. Provide one of these balls to the queen. The queen will hopefully begin laying eggs on this ball. Be sure to place the pollen ball in an area you can access, so you can easily monitor it. We place our pollen ball on a small plastic dish (a 2 x 2 inch plastic weighing dish available through lab supply stores). The pollen will become dry and hard within a few days. Replace it every few days unless there is evidence of eggs having been laid on the pollen ball. This "evidence" will look like a small wax lump on the pollen ball.

Alternatively, you can coat the pollen ball by dipping it in melted beeswax. Keep the wax at as low a temperature as possible to prevent harming the nutritional content of the pollen. The wax coating will help prevent the pollen from drying. The coating also seems to add to the attractiveness of the pollen ball as a place for the queen to lay eggs. Wax-coated pollen balls will need to be replaced weekly if they are not being used.

Photo: Michael J. Traynor

POLLEN GRUEL

Fresh pollen should be mixed with just enough nectar solution to make a stiff dough.

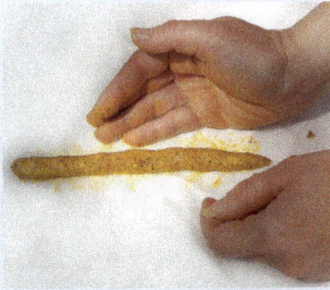

Photo: Michael J. Traynor

POLLEN CYLINDER

Pollen dough is rolled into a cylinder about 1/2 inch thick. Wax paper is an excellent working surface. To make pollen balls, cut the cylinder into 1/4 inch slices and form each into a little round ball.

Once the queen has laid eggs and they have hatched into developing larvae (preferably on the pollen ball) you will need to provide fresh pollen on a regular basis to support the developing brood. The cylinder of rolled-out pollen dough can be given directly to larger colonies that are being kept captive, or can be divided into halves or quarters for feeding developing colonies.

Growing colonies will need an amount of pollen equal to roughly one third of the size of the brood area each day. The brood area includes wax cells occupied by eggs, larvae and pupae. If there is a good deal left over the next day, reduce the amount of pollen you provide. The pollen cylinders can also be coated with wax if you want to extend time between feedings. The bees gradually uncover the wax, accessing the pollen which stays fresh under the wax seal.

Some species of bumble bees may prefer to receive pollen in the original loose pellet form. This tactic may work better with pocket-maker bumble bees, as these bees usually manipulate fresh pollen that has just been brought into the nest. Place the pollen in an open dish in your nest box. Again, the pollen will dry out quickly and may need daily replacement.

Photo: Michael J. Traynor

WAX COATED POLLEN

A small wax melter is used to keep wax at as low a temperature as possible without it solidifying again. Wax can be easily kept at an even temperature using a pan on an electric hotplate. Dip the pollen ball until it is evenly coated by the melted wax.

Bumble bees do not grow once they emerge from the pupal cocoon as an adult; the size of an adult bee is dependent on the amount and quality of food eaten as a larva. The individual bees within a bumble bee colony will typically vary greatly in body size. If the workers in your bumble bee colony are consistently tiny, you may not be providing them with enough pollen. Alternatively, there may be a problem with pollen quality or palatability.

However, if you provide too much pollen, your colony may begin queen production earlier than you would like. If queens are produced early in colony development, the working force of the colony will be reduced, because queens do little if any foraging for their natal colonies. You want to have a large worker population, both to support the colony by providing pollen to feed developing bees and to pollinate your crops, gardens and wildflowers.

Nectar

There are many different ways to provide access to the sugar solution that functions as a nectar substitute. Do not put an open dish of sugar solution in the nest box; your bees may accidentally drown in an open liquid source. Instead place the solution in some kind of dispenser.

It is best if the nectar container is outside of the nest so it can easily be monitored and replenished. Nectar can be dispensed at the top, side, or bottom of the colony. For dispensing from the top or side of the colony, use a bottle with a very small opening (some use a syringe), so that only one drop forms at a time. For dispensing through the floor, a feeder is set below the floor of the colony that the bees can access through a wick.

As we had trouble with leaking, sticky messes with the top and side feeders, we opted for the bottom feeder. With the bottom feeders, we are able to use larger containers, which means we do not have to spend a lot of time refilling small containers with sugar solution.

A sugar solution can be made by combining equal volumes of granulated sugar and water. To dissolve the sugar, heat the water, but be sure not to caramelize or burn the sugar. One way to avoid burning the sugar is to first boil the water, and then remove it from the heat source. Pour in an equal amount of granulated sugar and stir until the sugar is dissolved. You will be able to see the granules until they dissolve. Once the water has cooled

NECTAR SUBSTITUTE

9 cups water mixed with
9 cups sugar
Bring water to a boil.
Remove from heat. Add
sugar and stir until dissolved.

Refrigerate finished solution.

a little so that it will not burn your skin, test by rubbing a few drops of the solution between your fingers. If you can still feel grittiness, keep stirring.

Some people use honey water, which is closer to the nectar that bumble bees drink naturally. However, this requires frequent replacement, as honey is prone to fermentation when mixed with water.

Photo: Ian Burns

NECTAR POTS

Bumble bees store nectar in wax pots, constructed in a wide variety of shapes and sizes. Nectar is also stored in used pupal cells built up with wax. The bees produce the wax themselves, then mold it into shape using their mandibles (mouth parts). When all the pots are filled, more are constructed. Some pots are tiny, only large enough to fit a bee's head. Others are large enough that a bee could bathe in the pot. Occasionally, a pot will be constructed that exceeds the others in beauty of form, such as the pot to the far left of this picture. Any potter would be jealous.

CHAPTER SEVEN

ENCOURAGE AND CARE FOR YOUR COLONIES

Your job as a bumble bee rearer is not over once you place your queen in her housing along with food and warmth. You must be vigilant, keeping an eye on the queens to make sure they have access to the pollen and nectar you have provided, checking for any signs of eggs and helping to keep things clean.

Broodiness

In the starter box, there are several signs that egg-laying is imminent. The queen will be calm and spend most of her time in the nest area, as opposed to climbing the walls. She will eat some pollen, as evidenced by pollen-colored feces. She will deposit wax around the nesting area, sometimes constructing a cup to hold nectar.

A broody queen will sit on top of the pollen ball, hugging her body flat against it. This posture means she is incubating. If she has laid eggs somewhere other than the pollen ball, you will see her incubating a wax clump.

A broody queen will be particularly agitated when disturbed. If she shows no signs of broodiness in two weeks, release her. She is unlikely to initiate a nest in captivity. This is a fairly common occurrence. You may have to release half your queens due to lack of broodiness.

There are several different methods to encourage the queen bee to begin laying eggs. You can try putting two bumble bee queens into the same box, separating them after one has laid eggs. You could also try putting honey bee workers in with the bumble bee queen. Or, if you have other bumble bee colonies already established, you can place several newly emerged workers or some pupae in with the bumble bee queen. More elaborate, but highly successful, is to place a heated, fake pupa in the nesting box to encourage the queen to start laying eggs.

We also found that the queen sometimes responds positively to the presence of wax. We had better success rates when we gave the queen a pollen ball that was coated with beeswax. This has the added benefit of preventing the pollen from drying out quickly, resulting in less work for you.

Promoting queen broodiness is an exciting topic, as there is still much to learn about what we can do to increase the success rates for bumble bee queens starting nests, particularly for species that are not commonly reared. See Rearing References for more details on what we do know. We encourage you to observe the bees closely and try new things. Let us know how things go.

Handling Bumble Bees

You will need to develop skills in handling bumble bees. There are times when you will need to transfer bees between boxes or you will need to retrieve escapees. A pair of blunt ended tweezers is the best tool for grabbing bees from the nest. With experience and patience, you can gently pick them up by the back legs without causing them any harm. It is also helpful to keep some small jars handy in the bumble bee room to trap any escaped bees.

If you use only red light in the rooms where you work with the bees, the bees are less likely to fly out of the nest. Bumble bees do not see the red spectrum, so a red light disturbs them less than the typical lights we use in our homes. You can either purchase red colored light bulbs, or mount a red filter over the light source. This type of light source is frequently used in photographic darkrooms.

Photo: Michael J. Traynor

HANDLING BEES

Blunt ended tweezers work extremely well for transferring a bumble bee between boxes. Be sure to grab the bee by the hind leg. Here a *Bombus fervidus* queen is being transferred.

Keeping Things Clean

Bees prefer to defecate outside the nest. When they are confined, they have no choice but to defecate in the nest. They will usually have one or two areas in the nest, typically corners or edges, where they will relieve themselves.

It can be difficult to keep the nest clean when the bees are confined to the nest. They will often choose a corner in which to defecate. If you put paper down in one of these areas, be sure to tape down all edges as the bees frequently tear the paper up and use it to cover the brood clump. In some of our colonies we have made a false floor out of hardware cloth. Most debris falls through to an area where it can be removed. The bees have no trouble walking on hardware cloth.

Some bumble bee breeders add a side room to the finishing box. This gives the males an area to congregate, as well as allowing the bumble bees to defecate away from the brood nest area. If you are reusing equipment, make sure to clean it with hot water and bleach to sterilize it. Otherwise you risk negatively affecting the health of your colony.

Free-flying Bees

Once your colonies begin producing workers, you can keep them confined and continue to provide them with all their pollen and nectar. Alternatively, you can open the nest to the outside, allowing them to forage. If you decide you want to let your bees forage there are several precautions you should take. Foraging is a risky endeavor. You are bound to lose some bees when you open the colony to the outside.
It is important to wait until the colony is strong, meaning there are at least twenty workers present, before letting them have access to the outside. Even then, the bees may require food supplementation until they get used to foraging. We recommend providing all colonies with nectar and pollen during at least the first week of their release. If the bumble bee species you are working with is a pollen-storer you can look into the wax cells to see if they have stored pollen. For all bumble bees, you can see if there is nectar in the nectar pots.

After the bees are allowed to forage freely, it is a good idea to monitor the nest occasionally to make sure they have enough food. Pollen can be placed right into the nest. If there are empty wax cells, nectar may be

THE NEXT GENERATION

Even if you don't notice the larger queen larvae in your colonies, you should notice the queen pupae, shown here on the right. Queen pupae are often two to three times as large as worker pupae. Queen production often signals the approach of the end of a colony's life cycle.

Photo: Elaine Evans

poured directly into these cells.

If you are placing the box directly outside, be sure to place it in the shade. Secure the lid so that it will not blow off in the wind. More protection may be needed if there are skunks or bears in the area, as bumble bee colonies are a tasty treat for them. You may also need to protect your colony from invaders, such as ants and wasps that will be attracted to the nectar source.

Another option is to keep colonies inside a building, with access to the outside. This is the best choice if you wish to observe the bees in their nest. Colonies can be set with their entrance right next to a hole in the wall or a tube can connect the nest and the hole in the wall. The tube should be big enough to allow two bees to fit in it at the same time, so that bees can exit and enter simultaneously. However, the entrance should not be much larger than this, to avoid predators taking advantage of this entrance.

Queen Mating and Hibernation

Some colonies will produce queens and males. This usually happens at the end of the colony's life cycle. Depending on the species you are rearing and conditions within the colony, the production of queens and males will vary, both in seasonal timing and quantity.

Watch your colony to see when queen production occurs. Signs to look for are enlarged cells containing queen pupae, and the production of males. It is possible to allow queens produced by your colony to mate in confinement with males from another colony and then store queens through the winter in a controlled environment. This procedure will provide you with fertilized queens at any time, without having to wait until spring to catch them. Year round production of bumble bee colonies depends on this year round access to queens.

Our experiences with mating and over-wintering queens are limited to *Bombus impatiens*, a bee that mates readily in small spaces. We used a 3' x 3' x 3' cage. Queens can be introduced to the cage several days after they emerge from their pupal cocoons. Introduce at least twice as many males as queens into the cage. To prevent inbreeding, the males must be from a different colony than the queen.

Some queens may mate with more than one male. *Bombus impatiens* queens mate readily in confinement and do not seem to have any special requirements. Other bees may need sunlight, may mate only at certain times of day, or may prefer a larger area for flying.

After mating, give the queens access to nectar and pollen in a cage similar in size to the mating cage located in a dark cool room. Multiple queens can be kept in the same cage. Wait for them to become less active, which should take a few days at most.

Put them in a small, clean container, like a matchbox or plastic vial, with some ventilation. Place them in a refrigerator with the temperature set between 30° and 40° F. To ensure the queens do not dry out, they need some humidity. Place

Photo: Ian Burns

BUMBLE BEE MATING

A male and queen *Bombus impatiens* are mating. The much smaller male grasps the queen and keeps a tight grip on her even after mating has ended, trying to keep other males from mating with the queen. The tenacious males can maintain their grip even with the queen flying, struggling with the extra weight.

43

a damp sponge in the refrigerator. Keep the sponge away from the bees otherwise you could inadvertently encourage mold. The queens can remain in the refrigerator through their normal hibernation period of four to eight months.

You can bring queens out of hibernation earlier also. Some people don't bother with hibernation at all, but take the queen right from mating to starting a nest. This usually involves anesthetizing the queens with carbon dioxide gas. Carbon dioxide narcosis seems to stimulate queens to begin laying eggs. For this you will need a container of carbon dioxide gas. Use a plastic tube to dispense gas into a vial containing the queen. Turn the gas off once the queen is knocked out. After she awakens, some people put the queen in a large cage with access to nectar and pollen to encourage her to fly. They then place the queen in a starter box. There is no need to knock out a queen after she has been hibernating for five months or more.

CHAPTER EIGHT

NOW GO HAVE SOME FUN!

Observing and collecting bumble bee queens on flowers in the spring is a very enjoyable pastime! Locating and capturing bumble bee queens can involve the entire family. It also makes a great project for youth groups. Everyone is sure to learn something new about nature and bees in particular.

We realize that these methods will require some trial and error for success. See the table in Chapter Eleven for a list of common problems and their solutions. We also understand that some people will prefer to have a bumble bee queen shipped directly to their doorstep, so they do not have to engage in the effort to collect them. However, you will be missing out on a lot of fun.

In raising your own bumble bee colonies, you will see firsthand how bumble bees work to care for their young, collect and store their food, and maintain their nests. There is always something happening within the busy bumble bee colonies.

This manual was written to help stem the disappearance of our bumble bees. We want to encourage the use of local bumble bees for pollination and support healthy bumble bee populations by providing appropriate nesting sites and food sources. You can help bumble bees by being good stewards of their colonies and propagating plants they pollinate.

The more people that understand the importance of bumble bees' role as pollinators, the more likely we are to find ways to support their well being, as they support our well being through providing us with the food their pollination services have produced. Share the knowledge you gain from your experiences with bumble bees with others. Together, we can make the world a friendlier place for our buzzing companions.

And Stay In Touch

You may discover techniques or design equipment that increase your success or efficiency as a bumble bee rearer. We encourage you to share your experiences with us as we work to increase ecosystem health and agricultural productivity through supporting healthy pollinator populations. Please stay in touch with us through our website (www. befriendingbumblebees.com) where we hope you will share your bumble bee experiences with us. We would love to hear about your success stories, for there are sure to be plenty!

BUMBLE BEE QUEEN FOUNDS A NEST

A *Bombus pensylvanicus* queen, with her pollen baskets full, searches for a place to start a new colony.

Photo: Doug Golick

PART THREE

ADDITIONAL
INFORMATION

CHAPTER NINE

NATURAL ENEMIES AND NEST ASSOCIATES

Other living things may locate your bumble bee colonies and try to take up residence. Don't let this frighten you. Many are scavengers and feed on pollen and nest material. Others live in or on the bumble bees themselves. Some are obvious and easily detectable, while others may only be recognized when bees die or colonies collapse. We describe some below in detail, so you can be on the lookout for the problematic species and ignore the benevolent residents. Since the art of bumble bee rearing is still relatively new, we are always learning more about protecting our colonies. Don't let this information discourage you.

To defend your colonies from marauders, inspect your colonies on a regular basis, remove unwanted strangers, keep your equipment clean and use local bumble bees. There are no chemical deterrents or medication to use for bumble bee pests and diseases. For the safety of the bees, the environment and your selves, do not experiment with your own pesticide applications.

Nest Demolishers: Wax Moths and Hive Beetles

The most obvious and common bumble bee pests are wax moths. There are several different species of wax moths (*Galleria mellonella, Aphomia sociella,* and *Vitula edmandsii*). Wax moth larvae chew though the bumble bee nest, leaving a trail of silk webbing and debris. These creatures can be very destructive to the physical structure of bumble bee nests. It is difficult to remove all of them once they are established, but vigilance and early removal can help deter an infestation.

Small hive beetles (*Aethina tumida*) have become an increasing problem for honey bees in the U.S. over the past several years causing damage to comb, stored honey and pollen. They are also able to infest bumble bee nests, destroying the nest and consuming pollen stores. Small hive beetles are currently most predominant in the southeastern U.S., but their range is increasing. Small hive beetles are easily detectable as they make quite a mess. Remove them from colonies as quickly as possible.

The Masqueraders: Psithyrus

Besides *Bombus* there is another group of bumble bees in North America called *Psithyrus* (pronounced "Sitherus"). *Psithyrus* are closely related to the more familiar *Bombus*. Although closely related, the two groups differ greatly in their habits.

Psithyrus are social parasites of *Bombus* colonies. Instead of founding their own nests, they take over nests that have already been started by *Bombus* queens. Sometimes the *Psithyrus* female will enter the *Bombus* colony and kill the queen. Other times the *Psithyrus* female will hide in the nest and lay eggs. Either way, the *Psithyrus* brood is fed by the workers of the colonies as if their proper queen had laid the eggs.

However, the *Psithyrus* larvae do not develop into workers. As adults, *Psithyrus* do no work to help support the colony. They have no pollen baskets on their legs to carry pollen back to the colonies. They live off the pollen and nectar collected by their hosts. A *Psithyrus* invasion eventually leads to a stop in worker production and colony decline.

If the *Psithyrus* females are seen before they kill the queen, they can be removed from the nest without having an impact on the colony. They look like other bumble bees but often have less hair on their abdomens and have no pollen basket on their hind legs. See Chapter 10 for more information and techniques to rear and study this group. See table in Chapter 11 for a list of *Psithyrus* species in North America.

Pests and Parasites: Wasps, Flies, and Mites

Less obvious bumble bee pests include parasitic wasps, such as Braconidae and Eulophidae, which lay their eggs on bumble bee larvae and pupae. The larvae of these wasps then feed on the bumble bees and kill them. *Melittobia* spp. (Eulophidae) are parasitic wasps that have proven to be a problem in some rearing facilities. Unfortunately the parasitic wasps are very small, sometimes measuring less than one-sixteenth of an inch in length, so keep a sharp eye out for these. If you have a colony suddenly collapse, isolate the colony as quickly as possible.

There are also parasitic flies (Conopidae and Phoridae) that lay their eggs on the abdomens of adult bees while these bees are out foraging.

The larvae of Conopid flies develop within the abdomen of the adult bee, killing them after several weeks. If you have free-flying colonies there is little you can do to control this parasite, as the bees are parasitized while they are out foraging. Since the foragers can live for several more weeks even after they have been parasitized by Conopid flies, this parasite is not as potentially destructive to bumble bee colonies as those that attack the young. However, parasitization by Phoridae flies poses a more dire threat as the bees usually die within five days. Foraging is risky business, which is why we recommend waiting until colonies are at least twenty bees strong before opening the door to let them forage freely.

Sarcophagid fly larvae (*Brachicoma* spp.) attack *Bombus* larvae in the nest, thus slowing colony development. If there are enough *Brachicoma* larvae present they can prevent queen production in colonies. Keep an eye out for this pest and, if detected, isolate colonies and manually remove the larvae.

You may have heard of the mites wreaking havoc in honey bee colonies. Thankfully the mites that take up residence in bumble bee colonies are of a different variety, and do not seem to cause as much of a problem as the more destructive honey bee mites.

Some mites (*Parasitus* spp.) are external and can be seen with the naked eye. Although common on the backs of queens in the spring, the effects of these external mites are not known. Some suggest removing mites from the queens before rearing attempts are made to err on the side of safety. These external mites appear to be scavengers, feeding on pollen after gaining entrance to the nest by hitchhiking on the queens.

Others mites are internal and are very small, requiring dissection and microscopic examination for detection and identification (*Locustacarus buchneri*). These mites feed on the blood of the bees and appear to be involved in virus transmission. If you notice deformed wings among your new bees, this may be a symptom of a virus that was spread by internal mites.

The impact of these internal mites is not currently well understood. While the mites appear to shorten the life span of adult bees, they do not appear to pose a great threat to colony health. However, if the mites transmit a more virulent virus, or prey on a more vulnerable host, the threat could be a serious one.

The Others: Diseases, Fungi, and Other Attackers

Nosema bombi is a microsporidian (single celled organism) that infects
The digestive tract of bumble bees. A *Nosema* infection can destroy a
bumble bee colony, and it has been a problem in some commercial rearing
facilities. *Nosema bombi* is spread through contact with feces. If your
colonies are open to the outside, keep the colonies separated, so that they
remain isolated in case of an infection, as the disease can spread quickly.
Be sure to sterilize all equipment.

Other single celled organisms that may cause problems in bumble bee
colonies include *Crithidia bombi* and *Apicystis bombi*. It is important
that you quickly remove any colonies that decline rapidly and sterilize
equipment before reusing it.

There are also many fungi that attack bumble bees at different life stages.
Little is currently known about these fungi, but they may appear as
discolorations on larvae or pupae. Be sure to isolate any colony where you
see sign of fungi and sterilize your equipment.

Problems for Queens

The pests and diseases mentioned above primarily affect workers and
brood. There are several diseases that primarily affect spring queens.
Queens that die within a few days of capture are often infected by the
disease *Apicystis bombi*. Others that live for several weeks, but never
start laying eggs, are often infected by *Sphaerularia bombi*, a roundworm
(nematode) that lives in the abdomen of bumble bee queens, frequently
preventing queens from founding nests. Mermithrid nematodes can also
parasitize bumble bees.

Infection and parasitization by these organisms may be the explanation
for why many captured queens do not begin laying eggs, so do not blame
yourself if a particular queen refuses to lay eggs.

The Good and the Harmless

Not all visitors to bumble bee colonies are destructive. Some merely scavenge
amongst the debris. Carpet beetles (Dermestidae), and flour beetles (*Tribolium*
spp.) are some of the insects that are found in bumble bee nests feeding on

Several moth species, including Indian meal moths (*Plodia interpunctella*) and Mediterranean flour moths (*Ephestia kuehniella*) can invade a colony and feed on food reserves in a bumble bee nest. These usually only cause problems for previously weakened colonies.

There are beneficial parasitic braconid wasps (*Apanteles* spp.) which can attack moths infesting bumble bee nests, so not all tiny wasps you see in a bumble bee colony are automatically bad.

What to Do?

When you see a creature other than a bumble bee within your bumble bee nest, do your best to capture this creature. If it is an insect, locate an entomologist who can correctly identify it for you. Only after you determine the identity of the creature, can you effectively protect your colonies.

However, common sense offers a lot of protection too. If you are sharing equipment between colonies, make sure it is clean before bringing it in contact with bumble bees. If you are going to move brood or workers between colonies, first observe them closely to make sure they both come from healthy colonies. Finally, isolate colonies if there are drastic population drops.

See Bumble Bee Biology References, particularly Alford (1975) and Macfarlane et al (1995), and Strange et al (2022) for more information on bumble bee diseases, pests, and parasites. The digestive tract of bumble bees. A *Nosema* infection can destroy a bumble bee colony, and it has been a problem in some commercial rearing facilities. *Nosema bombi* is spread through contact with feces. If your colonies are open to the outside, keep the colonies separated, so that they remain isolated in case of an infection, as the disease can spread quickly. Be sure to sterilize all equipment.

Other single celled organisms that may cause problems in bumble bee colonies include *Crithidia bombi* and *Apicystis bombi*. It is important that you quickly remove any colonies that decline rapidly and sterilize equipment before reusing it.

There are also many fungi that attack bumble bees at different life stages. Little is currently known about these fungi, but they may appear as discolorations on larvae or pupae. Be sure to isolate any colony where you see sign of fungi and sterilize your equipment.

Problems for Queens

The pests and diseases mentioned above primarily affect workers and brood. There are several diseases that primarily affect spring queens. Queens that die within a few days of capture are often infected by the disease *Apicystis bombi*. Others that live for several weeks, but never start laying eggs, are often infected by *Spaerularia bombi*, a roundworm (nematode) that lives in the abdomen of bumble bee queens, frequently preventing queens from founding nests. Mermithrid nematodes can also parasitize bumble bees.

Infection and parasitization by these organisms may be the explanation for why many captured queens do not begin laying eggs, so do not blame yourself if a particular queen refuses to lay eggs.

The Good and the Harmless

Not all visitors to bumble bee colonies are destructive. Some merely scavenge amongst the debris. Carpet beetles (Dermestidae), and flour beetles (*Tribolium* spp.) are some of the insects that are found in bumble bee nests feeding on pollen and other nest debris on the abdomens of adult bees while these bees are out foraging.

CHAPTER TEN

RAISING CUCKOO BUMBLE BEES

Very little is known about interactions between cuckoo bumble bees in the subgenus *Psithyrus* and their bumble bee hosts. This lack of knowledge is mostly due to the rarity of cuckoo bumble bees in space and time but also because continuous observations of wild nests are nearly impossible. Rearing them is thus a very useful way to better understand them. It is important that we start looking at them more closely as they are at even greater risk of extinction than many other bumble bees given their rarity and reliance on host bumble bees.

Cuckoo bumble bee biology
(subgenus *Psithyrus*)

Making a nest, feeding, and defending offspring takes a lot of energy for a bumble bee queen. Cuckoo bumble bees, also called *Psithyrus* (pronounced "Sitherus"), manage to avoid these tasks by tricking other bumble bee species into doing it for them. *Psithyrus* females lack the pollen baskets needed to carry pollen and cannot produce their own workers. They are entirely reliant on their hosts to feed their offspring. However, to do so they cannot just dump their eggs in another nest, like cuckoo birds do. Cuckoo bumble bees have to trick the entire colony and infiltrate their society to force host workers into rearing cuckoo brood.

Psithyrus have several challenges to overcome to complete their life cycle. They must 1) locate their host, 2) infiltrate the host colony, 3) gain acceptance as a colony member, and 4) take over control of the colony by replacing the host queen. Different strategies are used by cuckoo bumble bees to overcome each of these hurdles.

Host nest location and selection

Cuckoo females locate an established nest by smell, finding the entrance to the host nest following scent trails left by workers. Once the nest is located, the *Psithyrus* females first assess the size of the host colony. If it is too big, the intruder will be quickly killed or rejected by the workers. If

Cuckoo nest invasion

Fight between host queen (right)
and cuckoo female (left)

Host queen killed
by cuckoo female

Cuckoo female protecting her
eggs

Emergence of new cuckoo
females

Photos: Patrick Lhomme

the colony is too small, there will not be enough workers to care for future parasitic larvae. Cuckoo females must therefore be very choosy.

Host nest infiltration

Once the nest is chosen, cuckoo females must overcome the barrier of recognition set up by the hosts. At the time of intrusion into a host colony, *Psithyrus* females are generally recognized as alien and attacked. When they enter the nest, it's common to see several host workers mass around the *Psithyrus* female, each trying to bite and sting her. Despite this, death of the *Psithyrus* female is rare because they are heavily armored with larger and stronger mandibles, a hardened exoskeleton, and a more powerful sting. Often several of the attacking workers are killed by the invading *Psithyrus* female. However, most often the *Psithyrus* females try to avoid being attacked, either by avoiding contact with their hosts and hiding under the brood until they acquire the odor of the colony, or, more rarely, using a repellent chemical to prevent worker attack.

Integration into the host colony

Psithyrus females have evolved different strategies to overcome host recognition and integrate themselves into the host colony. Some species use a strategy of chemical invisibility by producing no odor to remain undetectable. The female then acquires host odor passively through contact with nest material and members of the colony. Such chemical camouflage also suppresses or reduces worker hostility. Some species can produce the recognition signals of their hosts rather than passively acquire host odor. The ability of the cuckoo female to mimic or acquire the odor of the nest seem to vary according to the host breadth of the *Psithyrus* species. The species parasitizing only one bumble bee host usually use chemical mimicry whereas species that usurp the nest of several species (see Table in Chapter 11) mostly use chemical camouflage.

Most of the time, once integrated into the nest, the cuckoo female eliminates the host queen. There is, however, evidence of prolonged cohabitation with the host queen for some species. Cases of peaceful coexistence between the host queen and the female parasite have been observed in the cuckoo *Bombus variabilis* hosted by *Bombus pensylvanicus*.

Dominance and parental control

Once accepted as a colony member, the cuckoo female usually tears apart the egg-cells of its host and feeds on the eggs and nest provisions. She also often ejects from the nest the very young host larvae. The female cuckoo seem to mainly destroy the brood of the host because she needs space to lay her own eggs and material to build the cells. Cuckoo females have reduced wax glands and can't produce enough wax to build egg-cells. However, they can build new wax cells from scratch using nest debris. Providing extra nest material from other nests or old nest can help reduce host egg destruction. Destruction of eggs and young larvae might also be a way to force the host workers to incubate the parasitic eggs. Egg cells of *Psithyrus* can contain three to four times more eggs than other bumble bees, with cuckoo eggs being longer and thinner to maximize the number of eggs per cell.

Since most *Psithyrus* kill the host queen, they must signal their dominant status to keep host workers under control. They prevent worker reproduction and aggressiveness by using a combination of aggressive behaviors and production of chemicals that inhibit worker fertility. The cuckoo females start to lay eggs just a few days after invasion. Since cuckoo bumble bees must either empty the larval cells made by the host queen or use nest wax debris to build their own, The egg-cells of the cuckoo female are easy to recognize as they are usually dark brown with a smooth surface. First, the *Psithyrus* females incubate, feed, and protect their own larvae. They will then spend most of their time moving from one egg cell to another pushing workers away and sometimes mauling host workers if they attempt to open the egg cells. However, once the cuckoo larvae hatch the cuckoo female starts to tolerate host workers presence and lets them feed her larvae. Parasitic offspring start to emerge approximately 30 days after a successful nest invasion.

Mating and hibernation

Newly produced *Psithyrus* females and males will leave the nest to find mates. Once mated, females will look for hibernation sites like other bumble bee species. No differences have been found in hibernation sites between parasitic and non-parasitic bumble bees. However, they hibernate much longer than their host as they usually enter hibernation earlier and emerge later than their host.

Raising cuckoo bumble bees

Set up the host colony first

You should first collect nest searching host queens during early spring to found the host colony (see Chapter Four). Once the first batch of workers have emerged, move the colony to the finishing box. It is best to make a finishing box with two chambers to provide the cuckoo female with a small vestibule to let her freely enter the nest and also escape if she needs. For that, you can place a cardboard or wooden wall inside the box with a connecting hole to separate the nest from the vestibule.

Get the *Psithyrus* female ready

Collect female cuckoo bumble bees later in the spring. Cuckoo females emerge from hibernation a few weeks after their host species, so host queens will have already established their nests by the time the cuckoo females have emerged. They are often found searching for host nests or feeding on flowers. Cuckoo females are fairly easy to recognize and catch in the field with their slower and less vigorous flight compared to other bumble bees.

Once collected, keep the females individually in starter boxes maintained in the same environment used to rear the bumble bee hosts. Feed them pollen and sugar-water for a few days or until host colonies are ready to be parasitized.

Although cuckoo females can't collect pollen, they do need to eat pollen to help their ovaries to mature. The sexual maturity of the cuckoo females can impact the invasion success and the type of interactions they exhibit with their hosts. If the cuckoo female is not sexually mature, she might not show the level of aggressiveness required to take control of the colony and could be quickly rejected or even killed. They can be kept confined in starter boxes up to three weeks waiting for the availability of host colonies.

Invasion procedure

Provide the right host species for the *Psithyrus* you collected (see table in Chapter 11 to know the hosts of each cuckoo species). *Psithyrus* females should always be introduced in young colonies containing only the first batch of workers or at least only young workers. Newly emerged workers are fully dominated by their queen and not very aggressive so they will not compete with the cuckoo female for reproduction. The number of workers before *Psithyrus* invasion should be about 5 to 15 workers. If the host colony is too large, there will be too many workers defending the nest and the cuckoo might be killed or rejected.

Introduce the *Psithyrus* female into the vestibule, outside the host nest. The cuckoo female usually first tries to avoid any contact with the host workers and hides in the nest comb. Some workers try to bite and sting the cuckoo female who often responds by killing the aggressive workers. Several workers are generally killed within a few hours after female introduction. The cuckoo female will usually be accepted in the host colony after one or two days, meaning that aggression from the host workers has stopped. If the cuckoo female is not accepted or is still trying to escape after this period, remove her from the nest.

To increase invasion success, you can also remove the host queen and add the cuckoo in the nest before the first workers have emerged. This way the cuckoo will have time to acquire the host nest odor and will easily be adopted by the newly emerged workers. *Psithyrus* female have the same ability to dominate workers as the real queen. However, it is preferable to remove the host queen to facilitate *Psithyrus* social integration, since workers that have already been dominated by their queen are less disposed to be controlled by the cuckoo. Of course, if the purpose is to better understand how cuckoo females take over control of the host nest, it is better to keep the host queen present in the colony to mimic natural conditions.

The amount of offspring a cuckoo female can have usually depends on the number of workers available in the host colony. The more workers, the more cuckoo offspring will be raised. However, if the host colony is too large, it is difficult for *Psithyrus* females to dominate all host workers and control their reproduction. If the colony becomes too populous after emergence of the successive batches of workers, it is better to remove the biggest and/or most aggressive workers that could compete with the cuckoo female for reproduction. This way you will maximize cuckoo female reproduction.

Psithyrus mating and hibernation

Psithytus mating and hibernation methods are similar to those used for bumble bees that are included in Chapter Seven. If the *Psithyrus* females have spent less than two months in hibernation, follow the recommendations for carbon dioxide narcosis and repeat the narcosis at 24 and 48 hours after emergence from hibernation. Personal observations of colonies invaded with narcosed cuckoo females and non-narcosed females showed similar reproductive success.

After emergence, remove cuckoo males and females from the host nests. Place males in a flight cage exposed to natural light and provided with sugar syrup. Isolate females in starter boxes and feed them with pollen and sugar syrup for a week. After that, place females and males in the flight. Ideally males should come from other colonies to prevent inbreeding and there should be approximately one female for five males in the cage. After successful mating, isolate the females and feed them with pollen and sugar syrup for another week. Like for other bumble bee species, placed them in small contains (matchbox or plastic vial), with a lightly moist paper towel to keep up humidity, and overwinter them in a cold room at 35-40° F. for a minimum of 2-4 months. Queens that overwinter in captivity are often difficult to activate. After hibernation, keep the cuckoo female at room temperature for 1-2 days with pollen and sugar syrup. Then transfer to a rearing room or room with temperatures above 70° F. Increased temperature alone does not always break diapause. The use of CO_2 narcosis helps to activate hibernated cuckoo females (See procedure in Chapter 7).

If want to read further details on the natural history and rearing of cuckoo bumble bees you can read Lhomme et al. (2013, 2019).

CHAPTER ELEVEN

TABLES

Bumble Bees of North America

The following tables list all North American bumble bee species. Tables are grouped as follows: bumble bee species amenable to rearing, species that are difficult to rear either due to behavioral or practical reasons, and parasitic species. Species in red are recommended for rearing. Successful rearing methods for recommended species have been documented.

Range: "E" is east of 100° W longitude, "W" is west of 100° W longitude, "N" is north of 40° N latitude, "S" is south of 40° N latitude. "Arctic" is north of 66° N latitude.

Pollen-feeding: the method used to feed larvae in the nest. "Storer" is pollen-storer. "Pocket" is pocket-maker.

Abundance: Bumble bee abundances vary greatly from year to year. These are general observations. "Common" means that within most of its range, this species is frequently seen. "Uncommon" means that though regularly seen, there are few individual observed. "In decline" means that there are fewer individuals seen now than there were in the past in significant parts of their range. "Extinct?" means that it has been several decades since any individuals have been seen. Bees that are common in some areas may be uncommon in other areas, particularly at the extremes of their ranges.

Emergence: "Early" is early spring. "Mid" is mid spring. "Late" is late spring and early summer. These ranges of emergence times will vary throughout North America depending on climate. For USDA plant hardiness zone 4, "Early" ranges from mid-April to mid-May, "Mid" ranges from mid-April to mid-May and "Late" ranges from early June to July.

Tables are compiled from information in Griffin 1997, Hobbs 1966, Heinrich 1979, Kearns and Thompson 2001, Laverty and Harder 1988, Plath 1934, Thorp 1983, and Williams et al. 2014.

Species amenable to rearing

Bombus species	Range	Feeding	Abundance	Emergence
affinis	E	Storer	In decline	Early
appositus	W	Pocket	Common	Early
bifarius	W	Storer	Common	Late
bimaculatus	E	Storer	Common	Early
caliginosus	W	Storer	Uncommon	Early
centralis	W	Storer	Common	Late
crotchii	SW	-	Common	Early
flavifrons	W	Storer	Common	Late
franklini	W	Storer	Extinct?	Early
fraternus	E + SW	Storer	In decline	Early
frigidus	NE + NW	Storer	Uncommon	Early
griseocollis	E + NW	Storer	Common	Early
huntii	W	Storer	Common	Mid
impatiens	E	Storer	Common	Mid
melanopygus	W	Storer	Common	Early
mixtus	W	Storer	-	Late
morrisoni	W	Storer	Common	Mid
nevadensis	E + W	Storer	-	Mid
occidentalis	W	Storer	In decline	Early
perplexus	E + NW	Storer	Uncommon	Mid
rufocinctus	E + W	Storer	Common	Late
sandersoni	E	Storer	Uncommon	-
sitkensis	W	Storer	Uncommon	Early
ternarius	NE	Storer	Common	Mid
terricola	E + W	Storer	In decline	Mid
vagans	E + NW	Storer	Common	Mid
vandykei	W	Storer	Common	Early
vosnesenskii	W	Storer	Common	Early

Bombus species in RED are recommended for rearing based on Plowright & Jay 1966, Strange 2010.

Species difficult to rear

Bombus species	Range	Feeding	Abundance	Emergence
auricomus	E + NW	Storer	Common	Late
balteatus	W	Pocket	Uncommon	Late
borealis	NE + NW	Pocket	Common	Late
californicus	W	Pocket	Rare	Early
distinguendus	Arctic	-	Uncommon	Mid
cryptarum	NW	-	-	Mid
fervidus	E + W	Pocket	In decline?	Mid
fraternus	E + SW	Storer	-	Early
frigidus	E + W	Storer	-	Early
jonellus	Arctic	Storer	Common	Mid
neoboreus	Arctic	Storer	Uncommon	Mid
nevadensis	E + W	Storer	-	Mid
pensylvanicus	E + W	Pocket	Uncommon	Mid
polaris	Arctic	Pocket	-	Late
sylvicola	W		-	Mid

Parasitic Bumble Bees

Parasitic species	Range	Bombus Hosts
ashtoni/ bohemicus	E + W	***affinis****, ***terricola****, *occidentalis, bimaculatus, vagans*
citrinus	E	***vagans****, ***impatiens****, *bimaculatus, nevadensis, griseocollis, pensylvanicus, terricola*
fernalde/ flavidus	E + W	*appositus, occidentalis, rufocintus, perplexus, fervidus*
hyperboreus	Arctic	*polaris**
insularis	E + W	*appositus**, *flavifrons**, *ternarius**, *nevadensis**, *fervidus**, *impatiens, huntii , bifarius , rufocintus, occidentalis, terricola, mixtus*
suckleyi	E+W	*occidentalis**, *terricola, rufocinctus, nevadensis, fervidus, appositus*
variabilis	E+W+S	***pensylvanicus****, *terricola, auricomus*

* = Cuckoo female observed breeding; Bold = Main hosts;
Grey = Potential hosts that have never been recorded in a nest.

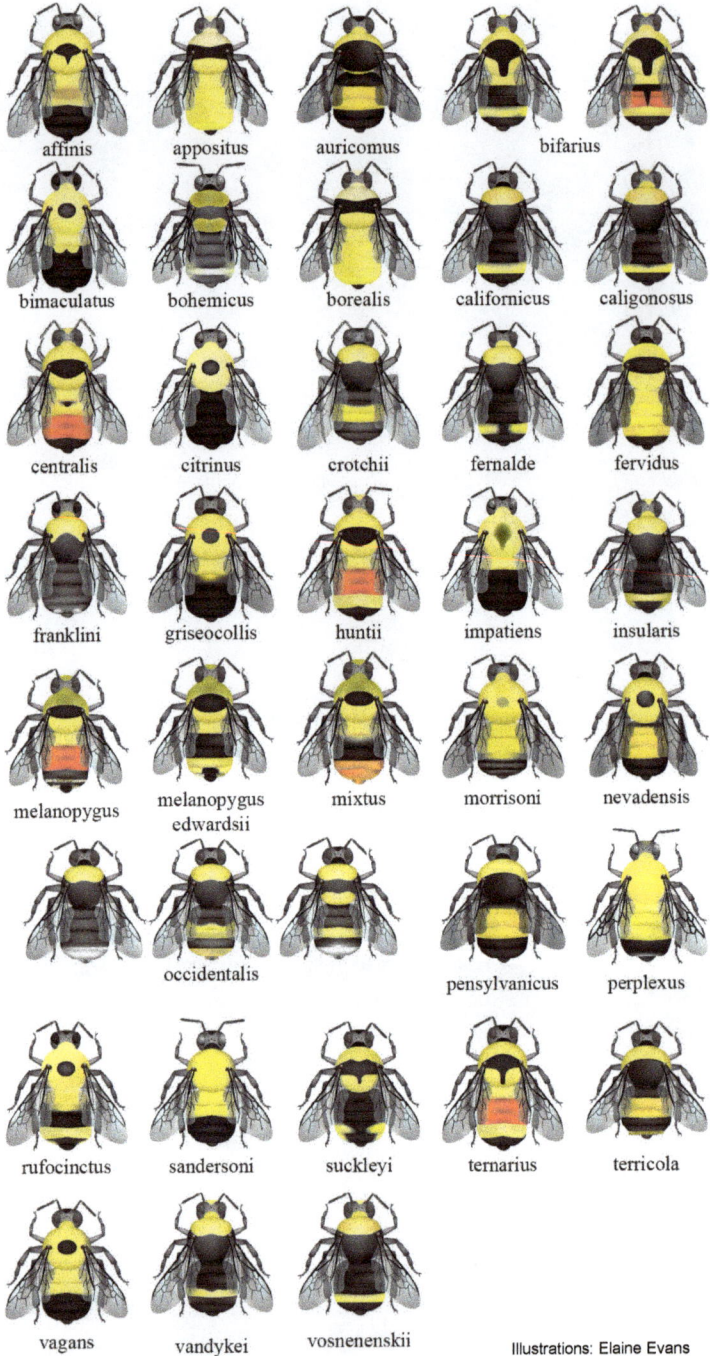

affinis   appositus   auricomus   bifarius

bimaculatus   bohemicus   borealis   californicus   caligonosus

centralis   citrinus   crotchii   fernalde   fervidus

franklini   griseocollis   huntii   impatiens   insularis

melanopygus   melanopygus edwardsii   mixtus   morrisoni   nevadensis

occidentalis   pensylvanicus   perplexus

rufocinctus   sandersoni   suckleyi   ternarius   terricola

vagans   vandykei   vosnenenskii

Illustrations: Elaine Evans

64

Troubleshooting

Problems	Solutions
tiny workers	provide more or better pollen
early queen production	provide less pollen
pollen dries quickly	coat pollen ball with beeswax
bees are not drinking	make sure feeder is not clogged
slow development of larvae	check room temperature, pollen quality
discolored larvae	isolate colony and sterilize equipment
sudden drop in worker numbers	isolate colony and sterilize equipment
pests within the nests	remove from colonies

Spring flowers for bumble bees

Family	Scientific name	Common name
Berberidaceae	*Berberis koreana*	Korean barberry
Boraginaceae	*Mertensia* spp.	Bluebells
Caprifoliaceae	*Diervilla* spp.	Bush honeysuckle
Caprifoliaceae	*Lonicera* spp.	Honeysuckle
Compositae	*Taraxacum officianalis*	Dandelion
Ericaceae	*Kalmia latifolia*	Mountain laurel
Ericaceae	*Pieris* spp.	Pieris
Ericaceae	*Rhododendron* spp.	Rhododendrons, Azaleas
Fabaceae	*Caragana aborescens*	Siberian pea
Hippocastanaceae	*Aesculus hippocastanum*	Horse chestnut
Hydrophyllaceae	*Hydrophyllum virginiana*	Virginia waterleaf
Iridaceae	*Crocus* spp.	Crocus
Laminaceae	*Mentha* spp.	Mint
Laminaceae	*Pycnanthemum* spp.	Mountain mint, Horse mint
Oleaceae	*Syringa* spp.	Lilacs
Rosaceae	*Malus*, *Prunus* spp.	Apples, Cherries
Salicaceae	*Salix* spp.	Willows
Saxifragaceae	*Ribes* spp.	Gooseberries
Scrophulariaceae	*Pedicularis* spp.	Lousewort
Violaceae	*Viola* spp.	Violets

These are flowers that are ideal focus points for your queen collection efforts. These are not recommended for plantings to encourage bumble bees as some of the plants are invasive.

GLOSSARY

Apidae: One of seven Families of bees in the Order Hymenoptera. Honey bees, bumble bees, carpenter bees, digger bees, cuckoo bees, orchid bees, and stingless bees belong to the family Apidae.

Bombus: All bumble bee species belong to this Genus.

Brood: The young, including eggs, larvae, and pupae.

Broodiness: Behavior indicating that a queen is tending eggs. The behavior includes eating pollen and holding her body close to the pollen ball to incubate eggs.

Buzz pollination: A behavior executed by a limited number of bees, including bumble bees, where they grab the flower and shake it by vibrating their wing muscles, releasing the pollen. This behavior is required by some flowers for successful pollination.

Carbon dioxide narcosis: CO_2 gas is used to stimulate queens to come out of hibernation and establish nests.

Caste: Members of a society that have distinct roles. In bumble bee societies there are three castes: queens, workers, and males.

Colony: A group of insects living together. With bumble bees, a colony is made up of a queen and her offspring.

Colony life cycle: Colony life cycle begins when a queen founds a nest by laying eggs. It continues with the growth period where workers are produced, and the reproductive phase with possible production of reproductives. Colony life cycle ends with colony decline, when the queen no longer lays eggs and colony population drops.

Diapause: An extended period of physical inactivity. Bumble bee queens normally enter diapause during the winter months.

Egg cell: A wax cup in which eggs are laid.

Finishing box: A box used to house bumble bee colonies after the first batch of workers have emerged as adults.

Hibernation: A period of inactivity in winter months.

Hymenoptera: One of thirty-two Orders in the Class Insecta. The Order Hymenoptera is comprised of ants, wasps, sawflies, and bees.

Larva: An early, limbless, stage of insect development. In bumble bees, larvae are sedentary. Remaining within the nest, the larvae are fed by workers and the queen. They grow underneath a wax covering that is expanded as they grow.

Males: Produced from unfertilized eggs. As such, they can be produced by both queens and workers. The primary function of males is to mate with virgin queens.

Mimicry: The resemblance of one organism to another. Batesian mimicry refers to mimicry in which a harmless species gains protection by mimicking a dangerous model species. There are many insects that mimic bumble bees in this way.

Mites: A highly diverse group of organisms related to ticks. Some live in soil or water, many are parasites living on or in other animals. Several mite species parasitize bumble bees.

Nectar pot: A wax pot constructed by bumble bees and used as a container for the nectar they gather from flowers.

Nest associates: Organisms that take advantage of shelter created by another organism.

Nest box: A box used to contain a bumble bee nest.

Parasites: Organisms that live in or on another organism at the expense of the host.

Pheromones: Chemical signals used for communication.

Pocket-maker: A bumble bee species using a method of feeding bumble bee larvae where foragers returning with pollen place it directly into pockets on the sides of wax clumps containing developing larvae. Pollen is not stored in separate wax pots.

Pollen basket: an indentation on the back legs of bumble bees used to carry pollen back to the nest. Gathered pollen is mixed with nectar and packed into the basket, or corbicula.

Pollen load: pollen packed into a bee's pollen basket.

Pollen pot: a wax pot used to store pollen.

Pollen-storer: a bumble bee species using a method of feeding bumble bee larvae in which pollen is stored in wax pots. Larvae are fed by the queen or workers, who peel back the wax covering the larvae and insert pollen.

Pollination: transfer of pollen from the male (anther) to the female (pistil) parts of plants, resulting in seed and fruit production.

Pupa: an intermediate, resting stage of insect development. During the pupal stage, bumble bees are restructured from their larval form into an adult. Larvae spin cocoons within which the transformation takes place. When they emerge from the cocoon, they are adults.

Queens: females responsible for most of the reproduction in bumble bee colonies. Queens are the only bumble bees capable of surviving the winter, and so are responsible for starting all the new bumble bee colonies each spring. Queens are larger than other females (workers).

Social insects: groups of related insects, living together with multiple generations present, and a division of labor.

Roundworms: a group of simple animals, many of which are parasitic.

Spermatheca: a specialized structure within female bees that functions as a sperm container.

Starter box: a box used to house early stages of bumble bee colony development, often with separate areas for feeding and for laying eggs.

Stinger: a needle shaped physical feature present at the end of the abdomen on some bees, wasps, and ants, which can deliver venom.

Workers: females responsible for colony maintenance, including gathering food, feeding young, cleaning, and defending the colony.

REFERENCES

BUMBLE BEE BIOLOGY

Alford, D. V. 1975. Bumble Bees. Davis-Poyner Limited, London.

Free, J. B., and Butler, C. G. 1959. Bumblebees. Collins Clear-type Press, London.

Goulson, D. 2003. Bumblebees: Their Behaviour and Ecology. Oxford University Press, Oxford.

Griffin, B. L. 1997. Humblebee Bumblebee: the Life Story of the Friendly Bumblebee and their Use by the Backyard Gardener. Knox Cellar Publishing, Bellingham, WA.

Heinrich. B. 1979, 2004. Bumblebee Economics. Harvard University Press, Cambridge, MA.

Kearns, C. A., and Thompson, J. D. 2001. The Natural History of Bumble Bees, a Sourcebook for Investigations. University Press of Colorado, Boulder, CO.

Macfarlane, R. P., Lipa, J. J., and Liu, H. J. 1995. Bumble bee pathogens and internal enemies. Bee World 76: 130-148.

Plath, O. E. 1934. Bumblebees and Their Ways. Macmillan Company, New York, New York.

Prys-Jones, O. E., and Corbet, S. A. 1987, 1991. Bumblebees. Cambridge University Press.

Sladen, F. W. L. 1912, 1989. The Humble-bee, its Life-history and How to Domesticate It. Macmillan & Co., London.

Strange, J. P., Colla, S., Duennes, M. Evans, E., Figueroa, L., Inouye, D., Lehmann, D., Moylett, H., Richardson, L., Sadd, B., Smith, J., Tripodi, A. Adams, L. 2022. Developing a Commercial Bumble Bee Clean Stock Certification Program: A white paper of the North American Pollinator Protection Campaign Bombus Task Force. https://www.pollinator.org/nappc/imperiled-bombus-conservation

REARING

Beekman, M., van Stratum, P., and Veerman, A. 1999. Selection for non-diapause in the bumblebee, *Bombus terrestris*, with notes on the effects of inbreeding. Entomologia Experimentalis et Applicata 93: 69- 75.

Christman, M. E., Spears, L., Koch, J. B. U., Lindsay, T. T., Strange, J. P., Barnes, C., and Ramirez, R. 2022. Captive rearing success and critical thermal maxima of *Bombus griseocollis* (Hymenoptera: Apidae): a candidate for commercialization?, Journal of Insect Science 22 (6)

Delaplane, K. S., and Mayer, D. F. 2000. Crop Pollination by Bees . CAB Publishing, CAB International, Wallingford.

Delaplane, K. S. 1995. Bumble beekeeping: the queen starter box. American Bee Journal 135: 743-745.

Delaplane, K. S. 1996a. Bumble beekeeping: inducing queens to nest in captivity. American Bee Journal 136: 42-43.

Delaplane, K. S. 1996b. Bumble beekeeping: handling mature colonies, mating queens. American Bee Journal 136: 105-106.

Delaplane, K. S. 1996c. Bumble beekeeping: second-generation queens. American Bee Journal 136: 439.

Griffin, R. P., Macfarlane, R. P., and van den Ende, H. J. 1991. Rearing and domestication of long tongued bumble bees in New Zealand. Acta Horticulturae: 149-153.

Lehmann DM. 2022. Protocol for initiating and monitoring bumble bee microcolonies with *Bombus impatiens* (Hymenoptera: Apidae). Bio-Protocol. Jun 20;12(12): e4451.

Lhomme, P., Sramkova, A., Kreuter, K., Lecocq, T., Rasmont, P., Ayasse, M. 2013. A method for year-round rearing of cuckoo bumblebees (Hymenoptera: Apoidea: *Bombus* subgenus *Psithyrus*). Annales de La Société Entomologique de France, 117–125.

Medler, J. T. 1958. Principles and methods for the utilization of bumblebees in cross pollination of crops. In 10th Internatl. Cong. Ent. Proc., Montreal, Aug. 17-23: 973-981.

Plowright, R. C., and Jay, S. C. 1966. Rearing bumble bee colonies in captivity. Journal of Apicultural Research 5: 155-165.

Pomeroy, N., and Plowright, R. C. 1980. Maintenance of bumble bee colonies in observation hives (Hymenoptera: Apidae). Canadian Entomologist 112: 321-326.

Pouvreau, A. 1993. Recherches sur les bourdons. Apidologie 24: 448- 449.

Ptácek, V. 1991. Trials to rear bumble bees. In 6th Pollination Symposium. Acta Horticulturae: 144-148.

Ptácek, V. et al. 2000. The two-queen cascade method as an alternative technique for starting bumble bee (*Bombus*, Hymenoptera: Apidae) colonies in laboratory conditions: a preliminary study. Psczelnicze Zaexyty Naukowe 44: 305-309.

Röseler, P. F. 1985. A technique for year-round rearing of *Bombus terrestris* (Apidae, Bombini) colonies in captivity. Apidologie 16: 165-170.

Strange, J. P. 2010. Nest initiation in three North American bumble bees (*Bombus*): gyne number and presence of honey bee workers influence establishment success and colony size. Journal of Insect Science 130.

Tasei, J. N. 1994. Effect of different narcosis procedures on initiating oviposition of pre-diapausing *Bombus terrestris* queens. Entomologia Experimentalis et Applicata 72: 273-279.

Tasei, J. N., and Aupinel, P. 1994. Effect of photoperiodic regimes on the oviposition of artificially overwintered *Bombus terrestris* L. queens and the production of sexuals. Journal of Apicultural Research 33: 27-33.

Treanore, E., Barie, K,. Derstine, N., Gadebusch, K., Orlova, M., Porter, M., Purnell, F., Amsalem, E. 2021. Optimizing laboratory rearing of a key pollinator, *Bombus impatiens*. Insects. 12(8):673.

Velthuis, H. H. W., and van Doorn, A. 2006. A century of advances in bumblebee domestication and the economic and environmental aspects of its commercialization for pollination. Apidologie 37: 421-451.

NEST BOX DESIGNS

Intenthron, M., and Gerrard, J. 1999. Making nests for bumble bees. International Bee Research Association.

Kearns, C. A., and Thompson, J. D. 2001. The natural history of bumble bees, a sourcebook for investigations. University Press of Colorado, Boulder, CO.

Munn, P. 1998. Helping bumble bees with *Bombus* nest boxes. Bee World 79: 44-49.

Plowright, R. C. and Jay, S. C. 1966. Rearing bumble bee colonies in captivity. Journal of Apicultural Research 5: 155-165.

Strange, J. P. Raising Bumble Bees at Home: A Guide to Getting Started. USDA-ARS, Pollinating Insect Research Unit. Logan, UT

BUMBLE BEE IDENTIFICATION FOR NORTH AMERICA

Golick, D. A., and Ellis, M. D. 2000. Bumble Boosters: A Guide to the Identification of Nebraska Bumble Bee Species. University of Nebraska Press.

Koch, J. et al. 2012 Bumble Bees of the Western United States. United States Forest Service. https://www.fs.fed.us/wildflowers/pollinators/documents/BumbleBeeGuideWestern2012.pdf

Laverty, T. M., and Harder, L. D. 1988. The bumble bees of eastern Canada. Canadian Entomologist 120: 965-987.

Stephen, W. P. 1957. Bumble bees of western America. Technical Bulletin of the Agricultural Experiment Station, Oregon State College, Corvallis. Bulletin 40.

Thorp, R. W., Horning Jr., D. S., and Dunning, L. L. 1983. Bumble bees and cuckoo bumble bees of California. Bulletin of the California Insect Survey. Volume 23: 1-79.

Williams, P. H., et al. 2011. Bumble Bees of the Eastern United States. https://www.fs.fed.us/wildflowers/pollinators/documents/BumbleBeeGuideEast2011.pdf

Williams, P. H., et al. 2014. Bumble Bees of North America: An Identification Guide. Princeton (New Jersey): Princeton University Press.

ABOUT THE AUTHORS

Elaine Evans is a University of Minnesota Extension Educator working on pollinator education and research relating to bee conservation. She completed her MS and PhD in Entomology at the University of Minnesota. Elaine studied declining North American bumble bee populations for the Xerces Society for Invertebrate Conservation. Her current work focuses on determining the status of bees in MN, monitoring threatened populations of the rusty patched bumble bee, evaluating the impacts of pollinator habitat on bumble bee colony performance and health, documenting pathogen spillover from managed honey bees to bumble bees, and enacting pollinator conservation through education, outreach, and public participation in science. Elaine works with members of the public who want to help bumble bees by joining one of the many regional Bumble Bee Atlases across North America.

Ian Burns completed his PhD studying social development and conflict in the North American bumblebee *Bombus impatiens*. During the course of his thesis work, he worked out many details of how to raise *Bombus impatiens* colonies from wild-caught queens. He is also a Chinese scholar and taught Chinese at the University of Minnesota as well as schools in Minneapolis and St. Paul. Ian is currently a part of University of Minnesota Bee Squad.

Patrick Lhomme is a researcher at the University of Mons in Belgium where his main interest is the taxonomy, ecology, and conservation of wild bees with a particular focus on cuckoo bumble bees. He studied the chemical ecology of cuckoo bumble bees and learned how to rear several European and American bumble bee species. He also developed more sustainable and biodiverse farming systems to protect wild pollinators and increase farmer income by improving crop pollination services with the International Center of Agricultural Research in the Dry Areas in Morocco and is the co-founder and head of R&D of Entomonutris, a company producing insect-based proteins for animal feed.

Marla Spivak is a MacArthur Fellow and McKnight Distinguished Professor in Entomology at the University of Minnesota. Recent awards include the 2015 Minnesota AgriGrowth Distinguished Service Award, the 2016 Siehl Prize laureate for excellence in agriculture, and the 2016 Wings WorldQuest Women of Discovery Earth Award. She and Gary Reuter bred a line of honey bees, the Minnesota Hygienic line, to defend against diseases and parasitic mites. Current current research includes studies of the benefits of propolis (tree resins) to the health and immune system of honey bees, the identification and biological activity of bee collected resins; the development of "bee lawns"—pollinator habitat in urban landscapes; the use of native forbs by honey bees through identification of collected pollen and decoding bees' dance language; the health of commercial honey bee colonies, the evaluation of queen breeders' efforts to select stocks for resistance to diseases and mite pests, novel methods to control Varroa mites in honey bee colonies; and surveys of native bees in Minnesota.

The University of Minnesota Extension service provides a wide range of opportunities to learn more about pollinators. The Bee Lab, through University of Minnesota Extension and the Department of Entomology, provides traditional classes in honey bee management as well as the innovative Bee Squad Program, which helps people foster healthy wild and managed bee populations and pollinator landscapes through education and hands-on mentorship. Bee Squad programs aim to discover creative ways to explore and support bees.

Learn more at beelab.umn.edu

INDEX

www.ingramcontent.com/pod-product-compliance
Lightning Source LLC
Chambersburg PA
CBHW040143270326
41928CB00023B/3338